FUNDAMENTAL CHANGES
IN CELLULAR BIOLOGY
IN THE 20th CENTURY

DE DIVERSIS ARTIBUS

<table>
<tr><td>COLLECTION DE TRAVAUX</td><td>COLLECTION OF STUDIES</td></tr>
<tr><td>DE L'ACADÉMIE INTERNATIONALE</td><td>FROM THE INTERNATIONAL ACADEMY</td></tr>
<tr><td>D'HISTOIRE DES SCIENCES</td><td>OF THE HISTORY OF SCIENCE</td></tr>
</table>

DIRECTION
EDITORS

| EMMANUEL | ROBERT |
| POULLE | HALLEUX |

TOME 43 (N.S. 6)

BREPOLS

PROCEEDINGS OF THE XXth INTERNATIONAL CONGRESS
OF HISTORY OF SCIENCE (Liège, 20-26 July 1997)

VOLUME III

FUNDAMENTAL CHANGES IN CELLULAR BIOLOGY IN THE 20th CENTURY

BIOLOGY OF DEVELOPMENT, CHEMISTRY AND PHYSICS IN THE LIFE SCIENCES

Edited by

Charles GALPERIN, Scott F. GILBERT and Brigitte HOPPE

BREPOLS

The XXth International Congress of History of Science was organized by the Belgian National Committee for Logic, History and Philosophy of Science with the support of :

ICSU
Ministère de la Politique scientifique
Académie Royale de Belgique
Koninklijke Academie van België
FNRS
FWO
Communauté française de Belgique
Région Wallonne
Service des Affaires culturelles de la Ville
    de Liège
Service de l'Enseignement de la Ville
    de Liège
Université de Liège
Comité Sluse asbl
Fédération du Tourisme de la Province
    de Liège
Collège Saint-Louis
Institut d'Enseignement supérieur
    "Les Rivageois"

Academic Press
Agora-Béranger
APRIL
Banque Nationale de Belgique
Carlson Wagonlit Travel -
    Incentive Travel House

Chambre de Commerce et d'Industrie
    de la Ville de Liège
Club liégeois des Exportateurs
Cockerill Sambre Group
Crédit Communal
Derouaux Ordina sprl
Disteel Cold s.a.
Etilux s.a.
Fabrimétal Liège - Luxembourg
Generale Bank n.v. -
    Générale de Banque s.a.
Interbrew
L'Espérance Commerciale
Maison de la Métallurgie et de l'Industrie
    de Liège
Office des Produits wallons
Peeters
Peket dè Houyeu
Petrofina
Rescolié
Sabena
SNCB
Société chimique Prayon Rupel
SPE Zone Sud
TEC Liège - Verviers
Vulcain Industries

D/1999/0095/57
ISBN 2-503-50888-X
Printed in the E.U. on acid-free paper

TABLE OF CONTENTS

Part one

ADAPTATION AND VARIATION OF CHEMISTRY AND PHYSICS
BY THE LIFE SCIENCES FROM THE 19th TO THE 20th CENTURY

Part two

THE NEW BIOLOGY OF DEVELOPMENT

PART ONE

ADAPTATION AND VARIATION OF CHEMISTRY AND PHYSICS BY THE LIFE SCIENCES FROM THE 19th TO THE 20th CENTURY

Foreword

Brigitte Hoppe

This volume presents a collection of selected papers worked out for a symposium at the XXth International Congress of History of Science in July 1997 at Liège, Belgium. The general aim of the group of papers presented at symposium 32 of the Congress was to analyze interrelations between the exact sciences chemistry and physics on one side and life sciences on the other side. It is well known that in many fields of biological sciences, mainly in those working with experimental methods, chemical and physical knowledge was integrated, but the historic development of that interrelation is not yet known and can not be explained enough in all details until the present day. By searching for the events in the past, by analyzing the former research programmes and the interpretations of results worked out with the help of experimental methods historians of science find out that introducing physical and chemical methods and knowledge into life sciences was not a simple but a very complex historic process. The reason was that the efforts to transform biology from descriptive natural history into an experimental " exact " science produced very deep changes not only of methods but also of thinking. Therefore, historians may find not only many unknown events of the development of life sciences from the 19th to the 20th century but they also have to analyze different elements and factors of the fundamental changes of biological thinking. Contributions to the understanding and explaining of important changes of biology are given by the papers of that volume presenting some detailed case studies on unknown events and theories as well as of their historic development.

The importance of physical and chemical methods became so effective in all fields of biology, that they belong to the topics of all papers presented, although methods are not explicitly mentioned. The impact and effects of immediate application of a new physical method, spectral analysis, in life sciences are discussed by J. Büttner. But the integration of fundamentals developed in the exact sciences at first arose also in a more indirect way : the life sciences adapted the general methodology and ideas concerning the aims of research and knowledge by modelling their research programmes. Between the middle of the 19th and that of the 20th century, many fields of the life sciences were dominated by a — more or less strict — reductionism which influenced the research programmes, interpretations of empirical data and phenomena, and theories. The authors themselves characterized them as based on mechanistic thinking. But these ideals were not adapted by all biologists without criticism.

Therefore, some papers discuss fields and theories based on strict reductionism like biophysics, while other papers describe research programmes and theories which were searching also for new interpretations trying to emphasize more the special organismic aspects of living objects. A case in this context was described by J. Janko : the development of research programmes and theoretical interpretations in the stimulation physiology in experimental botany from the 19th to the 20th century. A further research programme derived from the concept of Entwicklungsmechanik and developed in the early 20th century in Germany including many exchanges with the international scientific community is analyzed by B. Hoppe. In this paper the development of the then new field of experimental cytology working with the methods of " explantation " is described in detail. Not only the details of institutionalization of the branch in Germany and those of the first research results are presented, but also some implications of these researches leading the biologists to criticize the strict mechanistic thinking and to consider more special properties of the organic objects. A concept of physicalistic interpretation of living phenomena is discussed in the paper by R. Beyler. He describes unknown details of the early experimental researches in German radiobiology, introducing results of quantum physics into biology.

Further papers concerning the field of 20th century genetics and molecular biology discuss cases of the impact of organic chemistry and biochemistry in biological fields. The significance of modern chemical knowledge for the discovery of molecular aspects in modern genetics is reflected by the topics of the Nobel prizes for chemistry as is shown in the paper by C. Lentzner and A. Neubauer. C. Sinding presents aspects of the interactions between new knowledge of molecular biology, pharmaceutics and social behaviour of human individuals and the whole society.

Finally, the effects on theoretical biology of the transformation of life sciences from a " naturalistic " view to an evolutionary and a physicalistic interpretation of organisms are discussed by V. Gutina. The main consequences of different interpretations of biological phenomena for the creation of a general unifying " meta-theory " of 20th century biology are analyzed. But these discussions show also that some unresolved questions remain for future reasoning.

A REVOLUTION IN THE CLINICAL LABORATORY :
THE IMPACT OF THE INVENTION OF SPECTRUM ANALYSIS ON ANALYTICAL METHODOLOGY

Johannes BÜTTNER

INTRODUCTION

The emergence of " scientific medicine " in the 19th century brought with it characteristic changes in the ideas physicians held on the nature of diseases. Sickness was now interpreted as a disturbance of the physiological equilibrium in the sick organism, as a process which can be recognised with the aid of exactly defined symptoms or signs and can be causally explained. The more objectively such a sign can be determined, the more important it is for the diagnosis. Objective signs are in particular those which can be proved with the methods of the natural sciences. Another change was the inclusion of " chemical " signs, i.e. the results of chemical investigations, into the definition of a disease. A typical example is the disease with the cardinal symptom of albuminuria described by Richard Bright (1789-1858) in London in 1827[1]. However, to enable the course of the disease to be followed and therapeutical methods assessed, quantitative investigation methods are required and not only qualitative. Although such measuring methods were usual in physiology, the physician at the patient's bedside regarded them with great scepticism. More-over, no simple investigation methods were available. In the first half of the 19th century quantitative chemical analysis used to be made by separation of the substance to be measured followed by weighing. This procedure was not very useful for organic substances. Also the new method of elementary analysis, developed by Justus Liebig (1803-1873) to a reliable technique[2], needed sepa-ration and purification of the substances to be investigated. Then, from 1840

1. R. Bright, *Reports of Medical Cases, Selected with a View of Illustrating the Symptoms and Cure of Diseases by a Reference to Morbid Anatomy*, London, Longman, Rees, Orme, Brown and Green, vol. 1, 1827, vol. 2, 1831.
2. J. von Liebig, *Anleitung zur Analyse organischer Körper*, Braunschweig, Vieweg & Sohn, 1837.

onwards, due primarily to the work of Jean-Baptiste Biot (1774-1862)[3], the polarimeter became the first physical instrument to find its way into clinical laboratories. This instrument made it possible to determine quantitatively in very simple manner the glucose in the urine of patients with diabetes mellitus.

This was more or less the situation in clinical laboratories in 1860 when Gustav Robert Kirchhoff (1824-1887) and Robert Wilhelm Bunsen (1811-1899) developed spectrum analysis as a new method of chemical analysis. Below, the effects of this invention on the methodology in the clinical laboratory will be considered in somewhat more detail.

THE INVENTION OF SPECTRUM ANALYSIS

After a provisional communication in the autumn of 1859[4], Kirchhoff and Bunsen had published their method of spectrum analysis in two detailed papers of 1860[5] and 1861[6]. The studies related to the specific detection of elements in the emitted light of an adequately hot flame. The chemists became particularly clearly aware of the usefulness of their method when it led to the discovery of the two new elements Caesium and Rubidium. The papers contain an exact description of the spectroscope used[7]. As field of application of the method, apart from detecting elements " in earthly bodies ", the investigation of " heavenly bodies " was also expressly mentioned and reference made to work done by Kirchhoff on Fraunhofer lines in the solar spectrum[8]. There have been many descriptions of the long prehistory of spectrum analysis. Ever since the investigations of Joseph von Fraunhofer (1787-1826), spectra had been the subject of repeated investigations. Here we only wish to point out that it was not until the work of Kirchhoff and Bunsen that the attention of chemists was drawn to the use of spectrally split light for chemical analysis[9]. The exact description of the method and their spectroscope ideally suitable for its practical chemical application led to spectrum analysis establishing itself exceed-

3. [Jean Baptiste] Biot, " Mémoires sur la polarisation circulaire et sur ses applications à la chimie organique ", *Mémoires de la Classe des sciences mathématiques et physiques de l'Institut national de France*, [Paris], 13 (1835), 39-173 and 174-175.

4. G.R. Kirchhoff, " Über die Fraunhoferschen Linien ", *Monatsberichte der Königlich Preußischen Akademie der Wissenschaften, 1859* [Jahrgangszählung] (1859), 662-665.

5. G.R. Kirchhoff, R.W. Bunsen, " Chemische Analyse durch Spectralbeobachtungen ", *Annalen der Physik und Chemie*, hrsg. von J.C. Poggendorff, [Berlin], [Folge 2], 110 (1860), 161-189, with table V and VI.

6. G.R. Kirchhoff, R.W. Bunsen, " Chemische Analyse durch Spectralbeobachtungen : 2. Mitteilung ", *Annalen der Physik und Chemie*, [Leipzig] [Folge 2], 113 (1861), 337-382, with table III and V.

7. l.c. (6), p. 374-378 and table VII.

8. G.R. Kirchhoff, " Über das Verhältnis zwischen dem Emissionsvermögen und dem Absorptionsvermögen der Körper für Wärme und Licht ", *Annalen der Physik und Chemie*, hrsg. von J.C. Poggendorff, [Berlin], 109 (1860), 275-301.

9. See : M.A. Sutton, " Spectroscopy and the chemists : A neglected opportunity ? ", *Ambix*, 23 (1976), 16-26.

ingly rapidly in chemical laboratories. Only a short time after Kirchhoff's and Bunsen's publications, commercially produced spectroscopes had already appeared on the market. Most of these instruments were largely similar to the instrument used by Kirchhoff and Bunsen. The spectroscope made by the Munich instrument company Carl August von Steinheil (1801-1870), with whom Kirchhoff had co-operated, became widespread (fig. 1). This instrument also had an additional small deflection prism in front of the slit aperture and this made it possible to display a comparative spectrum in the field of vision.

THE DISCOVERY OF BLOOD SPECTRA

On its own, the rapid and reliable detection of elements was no reason for the physiological or pathological chemist to learn about the new method. A far more important aspect was that the new spectroscope devised by Kirchhoff and Bunsen made it very simple to investigate absorption spectra of coloured solutions as well. Felix Hoppe(-Seyler) (1825-1895), professor for applied chemistry at the University of Tübingen, illustrated this in 1862 with a blood solution. In solar light he observed the occurrence of two absorption bands[10]. Two years later the English physicist, George Gabriel Stokes (1819-1903), proved that Hoppe-Seyler had observed the spectrum of " arterial " haemoglobin charged with oxygen. By a special reducing agent (" Stokes's solution "[11]) Stokes was also able to illustrate the spectrum of oxygen-free " venous " haemoglobin[12] (fig. 2). These two discoveries immediately aroused the special interest of physiologists and physicians in spectrum analysis. With previous chemical methods, it had hardly been possible to determine the red colouring matter of blood and the numerous pigments in the animal and plant world[13]. Whether the red colour of blood was due to an inorganic iron compound or an organic dye had been the subject of discussion for many years. This question could now be answered by the spectrum analysis method. It was also possible to identify the red-coloured protein crystals which had been observed by Karl Bogislaus Reichert (1811-1883) as haemoglobin crystals. It now became possible to distinguish between haemoglobin and its derivatives and to follow directly the processes of the oxygen charging of the blood pigment. Chemical and physio-

10. F. Hoppe(-Seyler), " Ueber das Verhalten des Blutfarbstoffes im Spectrum des Sonnenlichtes ", *Archiv für pathologische Anatomie und Physiologie und klinische Medizin*, [Berlin], 23 (1862), 446-449.

11. Aqueous solution of $FeSO_4$ and tartaric acid added with ammoniumhydroxid until the resulting precipitate is dissolved.

12. G.G. Stokes, " On the reduction and oxidation of the colouring matter of the blood ", *Proceedings of the Royal Society*, [London], 13 (1864), 355-364.

13. Before 1860 the spectrum of chlorophyll, which can be isolated without great technical problems, has been reported by several authors (Brewster, Stokes, Angström, Harting), but these researches was not payed much attention by physiologists and physicians.

logical research on blood pigments now had at its disposal a method which quickly led to new discoveries of great importance to medicine.

IMMEDIATE INFLUENCES ON THE CLINICAL AND BIOLOGICAL LABORATORY

Even in the remaining years of that decade, that is the 1860s, the great interest in spectrum analysis led to a rapidly growing number of publications. The works on spectroscopy in pigment solutions, particularly those of biological origin, which are significant to our study started immediately after Hoppe-Seyler's and Stokes's publications on blood spectra. As early as 1863 we find a book by Gabriel Gustav Valentin (1810-1883), physiologist at the University of Berne, under the title " On the use of the spectroscope for physiological and medical purposes "[14]. Independently of Hoppe-Seyler, Valentin had observed the blood spectrum but did not publish his findings immediately. In his book he recommended spectrum analysis for the identification of biological pigments and, as did Hoppe-Seyler, for forensic detection of blood and toxic substances. The spectroscope was dealt with in textbooks on physiology from the middle of the 1860s. The detailed work of Ludwig Johann Wilhelm Thudichum (1829-1901), which appeared from 1867 onwards, was of particular importance. He was one of the early pioneers of spectroscopy of biological materials. Thudichum had studied medicine in Giessen and Heidelberg and had attended lectures by Bunsen and Liebig. In 1853 he emigrated to London. As " Chemist to the London Government Board " he began extensive investigations with the objection " of promoting an improved chemical identification of disease ". This was entirely consistent with the new concept of disease which we mentioned at the beginning. The starting point of his studies was to develop chemical tests in the cholera epidemic which broke out in London in 1854. In a report in 1867 he described in great detail his spectroscopic methods and his extremely well equipped Spectroscopic Observatory[15]. Of his numerous investigations of pigments, only his discovery of the luteine from Corpus luteum will be mentioned here ; this was the first carotene pigment[16] (fig. 3). In 1865 the spectrum microscope described by Henry Clifton Sorby (1826-1908) became of great importance in spectroscopic investigations in biology and medi-

14. G.G. Valentin, *Der Gebrauch des Spectroskopes zu physiologischen und ärztlichen Zwecken*, Leipzig, Heidelberg, C.F. Winter, 1863.

15. J.L.W. Thudichum, " Report on researches intended to promote an improved chemical identification of disease ", *Report of the Medical Officer of the Privy Council and Local Government*, [London], 10 (1867), Appendix 7, 152-294, with 6 coloured plates.

16. J.L.W. Thudichum, " Further report on researches intended to promote an improved chemical identification of disease ", *Report of the Medical Officer of the Privy Council and Local Government*, [London], 11 (1868), Appendix 6, 126-216, here p. 186 ; and " Über das Lutein und die Spectren gelbgefärbter organischer Substanzen ", *Centralblatt für die Medicinischen Wissenschaften*, [Berlin], 7 (1869), 1-5, here p. 2.

cine[17]. Although originally designed for petrographic studies, this instrument soon proved excellently suited to investigations of cells and tissues. The optician and astronomer John Browning (1835-1925)[18], marketed the Sorby instrument[19]. With this instrument, in the 1870s Sorby himself had performed extensive studies, in particular of vegetable objects. He attempted to establish a *Vegetable Chromatology*[20]. It should also be mentioned at this point that a committee of the British Association for the Advancement of Science, a committee appointed to investigate animal substances with the spectroscope[21], had formed around Sorby and the zoologist Edwin Ray Lankester (1847-1929) and had employed the spectroscope to investigate systematically numerous animal pigments[22]. Around 1870 the qualitative spectrum analysis of pigments had established itself as a method. It was employed in numerous biological and medical laboratories to identify and detect biologically, clinically and forensically important pigments.

QUANTITATIVE SPECTRUM PHOTOMETRY

The work of Kirchhoff and Bunsen on emission spectrum analysis, which led in physics and chemistry to astrophysical investigations and later to the study of atomic structures, gave rise to a completely different development in the field of biological and medical sciences. Here, initiated by the discovery of blood spectra, the centre of interest was absorption spectroscopy in solutions. Apart from detection of toxic substances the detection of elements played only a minor role. A factor which promoted this development was the excellent suitability of Kirchhoff's and Bunsen's spectroscope for investigating light absorption of pigments. The course of the further development led towards quantitative spectrum photometry. As we have shown at the beginning, in physiology and medicine there was a great need for quantitative analytical me-

17. H.C. Sorby, " On the application of spectrum-analysis to microscopical investigations, and especially to the detection of blood stains ", *Quarterly Journal of Science*, [London], 2 (1865), 198-214. Valentin had discussed in 1863 the possibility of a microscope-spectroscope in his book.

18. John Browning, oculist and astronomer, had opened a company for optical instruments in London. He produced among other instruments a handy " direct vision " pocket spectroscope of 3.5 inch of length which was widely used.

19. J. Browning, " On a simple form of micro-spectroscope ", *Monthly Microscopical Journal*, [London], 2 (1869), 65-66.

20. H.C. Sorby, " On comparative vegetable chromatology ", *Proceedings of the Royal Society*, [London], 21 (1873), 442-483.

21. Committee appointed to investigate animal substances with the spectroscope : E. Ray Lankester, " Report of a committee appointed to investigate animal substances with the spectroscope ", *Report of the British Association for the Advancement of Science*, [London], 38 (1868), 113-114 ; E. Ray Lankester, " Abstract of a report on the spectroscopic examination of certain animal substances ", *Journal of Anatomy and Physiology*, [London], 4 (1870), 119-129.

22. In this circle was also active the physician and biologist Charles Alexander MacMunn (1852-1911). He had reported in 1886 on the discovery of the histohämatines, which are today known as cytochromes. This discovery was forgotten after Hoppe-Seyler denoted them as artifacts.

thods. It is worth noting that physiologists were the first to attempt to develop quantitative methods for measuring the absorption of pigment solutions. The main interest was directed to a quantitative method for blood pigments. Even before the development of spectrum analysis, simple photometry of coloured solutions by visual comparison had been in common use[23]. This method was called *colorimetry*. The relationships between colour concentration, thickness of the traversed layer and light absorption were known[24]. Along with others, Bunsen had investigated these questions in photochemical work he performed with Henry Enfield Roscoe (1833-1915). But using a spectroscope to perform quantitative measurements was a new idea. In 1866 the Jena physiologist Wilhelm Preyer (1841-1897) described a method in which a blood solution in a glass cell in front of the spectroscope, initially allowing only red light to pass, is diluted with water until the first green can be seen in the spectrum[25]. By comparison with a haemoglobin solution of known content the haemoglobin concentration can be calculated from the dilution. From 1869 onwards, Karl Vierordt (1818-1884), physiologist in Tübingen, made a detailed study of quantitative spectrum photometry. It was based on sensory physiological investigations on the recognisability of brightness differences of coloured light by the human eye. In 1873, by a simple modification of the Kirchhoff-Bunsen spectroscope he succeeded in developing a practical method for spectrum photometry (fig. 4). Vierordt divided the entrance slit of the spectroscope into two sections and the width of these two sections was adjustable independently of each other using micrometer screws[26]. He allowed the unattenuated light of a petroleum lamp to enter through the upper section and in the front of the lower section he placed the pigment solution in a glass cell. The upper slit was then reduced using the micrometer screw until the eye perceived subjectively the same brightness in both slit sections. It was then possible to determine the pig-

23. The construction of a simple optical comparison instrument by the French optician Jules Duboscq (1817-1886) about 1850 made colorimetry easily applicable. Duboscq's immersion colorimeter has been used with small changes for more than hundred years in the laboratory. In the medical laboratory the Hämatinometer of Hoppe-Seyler was one of the first instruments for the quantitative determination of blood pigment (as haematin). See F. Hoppe(-Seyler), *Anleitung zur pathologisch-chemischen Analyse für Aerzte und Studirende*, Berlin, A. Hirschwald, 1858, 218-223.

24. " Law of Lambert and Beer " : Johann Heinrich Lambert (1728-1777) gave the first mathematical treatment of the attenuation of light passing an absorbing medium. See J[ohann] H[einrich] Lambert, *Photometria sive de mensura et gradibus luminis, colorum et umbrae*, Augusta Vindelicorum = Augsburg, E. Klett, 1760. See E. Anding [ed.], *Lambert's Photometrie. Zweites Heft* (Ostwalds Klassiker der exakten Wissenschaften, vol. 32), Leipzig, W. Engelmann, 1892, 66-67. August Beer (1825-1863) derived the relation between attenuation of monochromatic light and concentration of colour in solution. See [A.] Beer, " Bestimmung der Absorption des rothen Lichtes in farbigen Flüssigkeiten ", *Annalen der Physik und Chemie*, hrsg. von J.C. Poggendorff, [Berlin], [Folge 2], 86 (1852), 78-88 (with table I).

25. W. Preyer, " Quantitative Bestimmung des Farbstoffs im Blute durch das Spectrum ", *Annalen der Chemie und Pharmacie*, [Heidelberg], 140 (1866), 187-200.

26. K. Vierordt, " Über die Anwendung des Spectral-Apparates zur quantitativen Bestimmung von Farbstoffen ", *Berichte der Deutschen Chemischen Gesellschaft*, [Berlin], 4 (1871), 327-329 ; additions pp. 475, 519.

ment concentration from the light attenuation ratio. With this measuring arrangement it was possible for the first time to perform quite accurate determinations of the haemoglobin concentration. Further details of this development cannot be discussed here[27].

FURTHER DEVELOPMENTS AND PROSPECTS

Our short investigation of the first two decades following the discovery of spectrum analysis has made clear the enormous influence of the new method on medicine. The discovery revolutionised chemical analysis in the medical laboratory. In physics and pure chemistry the interest was focused on emission spectra and research on atomic structures by means of spectrum analysis whereas in medicine and biology the study of absorption spectra opened the way to the analysis of coloured biological materials.

In the clinical laboratory spectroscopy and quantitative spectrophotometry became standard methods for more than 100 years. The spectroscope made it possible for the first time to investigate blood pigment and its derivatives and to characterise and distinguish between a large number of pathologically important pigments. We have already got to know the quantitative spectrophotometry using the eye as a detector in its role as a development within physiology independent from basic science. The practical advantages of spectrum analysis for medicine were soon recognised. Compared with classical chemical analysis, the investigation methods were substantially simpler, more sensitive and could be performed more quickly. Methods were subsequently perfected for determining quantitatively not only pigments but also colourless substances. For this purpose reagents were employed which enter a colour reaction with the compound to be investigated. Finally, attempts were made to perform the entire analysis of blood and urine largely with spectrophotometric or colorimetric methods. Spectrophotometers and colorimeters are today still amongst the most important instruments in a medical laboratory. Their significance became even greater when, in the 1930s, objective measurements with photoelectric cells replaced the previous subjective measurement with the eye. After the Second World War these instruments formed the basis for the development of automatic analysis systems[28]. It is far easier to automate photometric measurements than it is to automate classical chemical analysis. Spectrum analysis is an interesting example of the effect of a physical discovery on biology and

27. Gustav Hüfner and independently Paul Glan described in 1877 improved spectrophotometers using polarization prisms instead of reducing the slit for light attenuation. Hüfners instrument became, after further improvements in 1889, the standard instrument for many decades. It was part of the equipment of many physiological and clinical laboratories in Germany.

28. The first automated instrument for clinical chemical analyses used a double beam filter photometer as measuring instrument. L.T. Skeggs, " An automated method for colorimetric analysis ", *American Journal of Clinical Pathology*, 28 (1957), 311-322.

medicine. The discovery of blood spectra made the possibilities of the new method immediately apparent to the physician. The further developments saw a remarkably intensive interaction between natural sciences and medicine. Quantitative absorption spectrophotometry, which was initially dependent on the human eye, had its roots in sensory physiological investigations in medicine. From this starting point, a multitude of detail inventions by physicists led to a comprehensive analytical system for the clinical laboratory. It was no doubt the practical advantages of these investigation methods which played the leading role in this development. Perhaps however the sensation of colour played its part. A physician is used to employing his senses for observations. The optical phenomena may have made his adoption of these methods easier, in contrast to other more abstract chemical methods. The ancient dream of physicians of employing the colour of urine as a diagnostic sign may also have had some influence.

FIGURES

1. Spectroscope of Kirchhoff and Bunsen (1861)

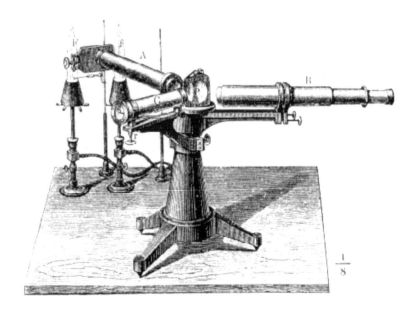

(From H.E. Roscoe, *Die Spectralanalyse in einer Reihe von sechs Vorlesungen,* 2. Aufl. Braunschweig, Viewejg, 1873, p. 69). This instrument was produced by C.A. Steinheil in Munich. A : collimator with slit, B : telescope, C : tube with a millimeter scale to be projected into the field of view.

2. Spectra of blood and derivatives

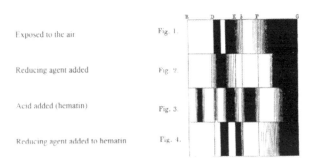

(From G. Stokes, *Proceedings of the Royal Society,* [London], 13 (1864), 356).
1. Oxygenated (" arterial ") haemoglobin, 2. " Venous " haemoglobin desoxy-
genated by Stokes' solution. 3. Haematin (Fe^{3+}-Protoporphyrin). 4. Haem
(Fe^{2+}-Protoporphyrin).

3. Spectrum of Luteine (= Carotene)

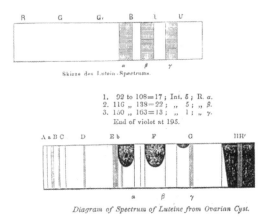

(Above from J.L.W. Thudichum, *Report of the Medical Officer of the Privy
Council and Local Government,* [London], 11 (1868), Appendix 6, 126-216,
here p. 186 ; below from J.L.W. Thudichum, *Centralblatt für die Medicinischen Wissenschaften,* 7 (1869), 1-5, here p. 2).

4. Spectrophotometer of Vierordt with double slit.

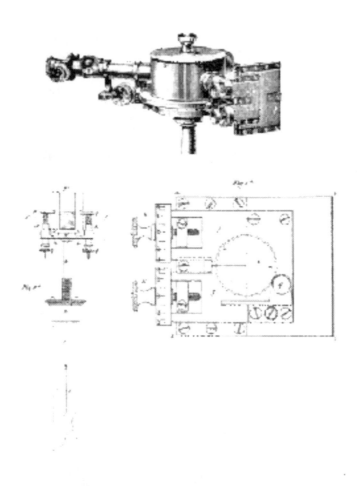

Above : instrument, to the right the double slit (from Gerhard Krüss und Hugo Krüss, *Kolorimetrie und Quantitative Spektralanalyse in ihrer Anwendung in der Chemie,* 2. Aufl. Hamburg, Leipzig, Leopold Voss, 1909, p. 123), below left : stand for the cuvette, right : detail of the double slit (from Karl Vierordt, *Die Anwendung des Spectralapparates zur Photometrie der Absorptionsspectren und zur quantitativen chemischen Analyse,* Tübingen, H. Laupp, 1873, table I, fig.1a and fig. 3b).

EXPERIMENTAL BOTANY, DEVELOPMENTAL MECHANICS AND *REIZPHYSIOLOGIE*

Jan JANKO

At the turn of the 19th and 20th centuries, experimental research in nearly all botanical disciplines was under the strong influence of the theoretical concepts mentioned in the title. Their application was supported by the methodological paradigm normally used in branches of zoology. However, it is hard to say which of these concepts was pivotal at the time. It is only a slight oversimplification to say that developmental mechanics prevailed in zoological disciplines, and the physiology of irritability (or, physiology of stimuli, *Reizphysiologie*) in botanical ones, but their dominance must be understood as gradual, not qualitative[1].

The article presented here aims to contribute to an elucidation of the interaction between the above mentioned theoretical frameworks used in experimental botanical research at the turn of the 19th and 20th centuries. At that time it was becoming increasingly clear that the mere fabrication of elaborately thought-out conceptual theories was not of itself sufficient. Experimentation designed to corroborate such conceptions often in fact showed up their irrelevance or placed them in a new light and into new and sometimes unexpected contexts.

It is interesting to note that the theoretical conceptions of the physiology of irritability in plants, for which W. Pfeffer in particular gained a reputation in his extensive work, were also elaborated with the help of the founder of developmental mechanics himself, W. Roux, who evidently wished his concepts,

---

1. On the general situation in biological sciences of this time see W. Coleman, *Biology in the Nineteenth Century*, New York, J. Wiley, 1971, VII + 187 p., and G. Allen, *Life Sciences in the Twentieth Century*, Cambridge (Mass.), etc., Cambridge University Press, 1978, repr. 1979. On the interplay between reductionistic and organismic concept at this time see B. Hoppe, " Chemophysiologie zwischen vitalistischer und mechanistischer Biologie im 19. Jahrhundert ", *Medizinhistorisches Journal*, 18 (1983), 163-183 ; J. Janko and S. Strbanova (eds), *Interrelations of Biology, Physics, and Chemistry in Historical Perspective*, Prague, Inst. for Theory and History of Sciences, 1991, 235 p.

originally meant for conditions prevailing in the animal kingdom, to gain the broadest application in the field of botany too[2]. Roux resented it when the younger generation of researchers took on terminology and methodology from a different source, as can be seen in his review of Küster's monograph on cecidia (1911)[3]. He reproached its author for accepting the concepts of K. Herbst, who conjectured that all external factors involve irritation, and put forward his own, as he saw it, more sophisticated division of external influences into releasing factors, differentiating factors, and irritation itself. Herbst himself, who was attempting to create a unified system of irritability connected to the developmental mechanics problem, also based himself principally on the zoological material with which he worked[4]. This made many botanists create new terms for their results or they avoided as much as they could the excessive theorizing for which German biologists in particular had a considerable penchant (compare H. Driesch, J. Reinke, J.V. Uexküll, R. Przibram, M. Kassowitz and others).

I have attempted to show that conceptions of developmental mechanics in experimental botany appeared in the form of five different, clearly defined research programmes : Schwendener's application of the principles of statics in research on plant stems and the arrangement of leaves (phyllotaxy) ; Sachs' conception of formative materials ; Voechting's concept of the decisive role of internal conditions for morphogenesis ; Klebs' programme of controlling plant forms by external conditions ; and finally Goebel's conception of so-called organography[5]. Out of these programmes, the ones that were significant for research on the boundaries between developmental mechanics and the physiology of irritability were primarily those of Sachs and Klebs. The copious research into the creation of cecidia, started under Sachs' inspiration by M. Beijerinck, who in 1888 spoke of " cecidogenic material "[6], comes under Sachs' conception, and was brought closer to the physiology of irritability issue when it was reformulated by E. Küster (1911). Sachs' interest in the issue of irritability is proved by his manuscript on periods of irritability written in the 1880s (I would like to thank Prof. H. Gimmler and Prof. W. Hartung from

2. On the branch of *Reizphysiologie* see, above all, the German secondary literature. Concerning Pfefferian physiology : H. Kniep, " Wilhelm Pfeffers Bedeutung für die Reizphysiologie ", *Naturwissenschaften*, 3 (1915), 124-129 ; E. Bünning, *Wilhelm Pfeffer : Apotheker, Chemiker, Botaniker, Physiologe 1845-1920*, Stuttgart, Wissenschaftliche Verlagsgesellschaft, 1975, 166 p.

3. E. Küster, *Die Gallen der Pflanzen*, Leipzig, S. Hirzel, 1911, 437 p. W. Roux' review in *Archiv der Entwicklungsmechanik der Organismen*, 35 (1913), 318.

4. C. Herbst, " Über die Bedeutung der Reizphysiologie für die kausale Auffassung von Vorgängen in der thierischen Ontogenese II ", *Biologisches Centralblatt*, 15 (1895), 721-745, 753-772, 792-805, 817-831, 849-855.

5. J. Janko, " Botanical Parallels to Developmental Physiology ", *Algorismus*, 21 (1997), 301-330.

6. M.W. Beyerinck (Beijerinck), " Über das Cecidium von Nematus Caprae auf Salix amygdalina ", *Botanische Zeitung*, 46 (1888), 1-11, 17-28.

Würzburg for the kind loan of the transcript) as by his later studies[7]. There was no other more penetrating experimental fusion of his research programme with the physiology of irritability during Sachs' lifetime — the renowned theory of tropism's expounded by J. Loeb (Sachs' pupil) actually only represented a translation of the *Reizphysiologie* concepts into the physiology of the nervous activity of animals and the physiology of movement in the spirit of extreme reductionism. In this spirit, tropisms were merely seen as manifestations of mechanical displacement (attraction and repulsion). As Ch.M. Child says, " the growth reactions of the plants leading to orientation with respect to external factors and known as tropisms constituted the starting point of Loeb's theory of tropism as the fundamental manner of reaction of the motile organism. The theory has been repeatedly and severely criticized and many of the criticisms remain unanswered "[8].

A fundamental breakthrough was achieved with Fitting's discovery of a substance in orchid pollen, which induced postfloration phenomena in the creation of the flower after pollination (1909)[9]. Fitting himself had worked on the classic subject of the physiology of irritability — inducement of tropic stimuli. Now he found that an aquatic extract of the active material from pollen inserted into the flower could induce premature decay. He called such artificially induced processes " induced postfloration processes " (*induzierte Postfloralvorgänge*). The " double effect " of orchid pollen had in fact been recognized as early as 1863 by F. Hildebrand[10]. In a theoretical commentary on his discovery, Fitting called the active material *Reizstoff* and classed it with hormones, i.e. the substances that had only recently been conceived by Starling. He attempted to apply the knowledge acquired from orchids to the shedding of the petals of geraniums and other plants, but this raised objections from Vöchting's student H. Wacker, and from Roux himself[11]. Both emphasized the endogenous nature of these phenomena as opposed to the externally determined character expounded by Fitting. Fitting could not resist to enrich biolog-

---

7. J. Sachs, *Bildungsreize und Bildungsperioden,* MS. (most probably from the sixties) in the Botanical Institute, Würzburg University. Compare his paper " Physiologische Notizen VII. Über Wachstumsperioden und Bildungsreize ", *Flora,* 77 (1893), 217-253.

8. Ch.M. Child, *Physiological Foundations of Behavior,* New York, H. Holt, 1924, 231. J. Loeb himself translated his plant physiology of stimuli into animal physiology of nervous activlty most spectacularly in his book *Einleitung in die vergleichende Gehirnphysiologie und vergleichende Psychologie mit besonderer Berücksichtigung der wirbellosen Tiere,* Leipzig, A. Barth, 1899, 207 p.

9. H. Fitting, " Die Beeinflussung der Orchideenblüten durch die Bestäubung und durch andere Umstände ", *Zeitschrift für Botanik,* 1 (1909), 1-86 ; " Weitere entwicklungsphysiologische Untersuchungen an Orchideenblüten ", *Ibid.,* 2 (1910), 225-266.

10. F. Hildebrand, " Die Fruchtbildung der Orchideen, ein Beweis für die doppelte Wirkung des Pollen ", *Botanische Zeitung,* 21 (1863), 329-333, 337-345.

11. H. Fitting, " Untersuchungen über die vorzeitige Entblätterung von Blüten ", *Jahrbücher für wissenschaftliche Botanik,* 49 (1911), 187-263 ; H. Wacker, " Physiologische und morphologische Untersuchungen über das Verblühen ". *Ibid.,* 522-578 ; W. Roux, " Über Cytochorismus ", *Ibid.,* 50 (1911), 355-356. Wacker avoided the term *Reiz* (stimulus) in his paper entirely.

ical terminology with a new term — chorism, a kind of irritation indicating induced reduction of concentration in tissue and cells. He explicitly designated his approach as *reiz- und entwicklungsphysiologisch*, i.e. connecting the physiology of irritability with the physiology of development. The active substance under consideration was of course not identical to the formative substance in Sachs' sense. Hildebrand's and Fitting's findings, like the cecidological research of other scholars were more in keeping with Herbst's conception of formative or organogenic stimuli (1901)[12].

According to Klebs, at the beginning of any process of organogenesis there must be irritation (*Reiz*), though in his later work he rather avoided this idea. One of the reasons for this disinclination on the part of Klebs and other experimenters was the terminological confusion and numerous disputes around the distinction between the concepts of irritation and release (*Auslösung*) propounded by the main representative of the physiology of irritability, W. Pfeffer[13]. Klebs thus represented an important stage on the way to the rejection of the concept of irritability by certain English, American and Dutch botanists at the beginning of this century, who considered it to be redundant and outmoded. They held that research carried out in a reductionist manner using the methods of chemistry and physics could well get by without it. Pfeffer himself preferred the concept of release as he admitted that the concept of irritability was often used in an unclear context. He attempted to emphasize the explanation of the phenomenon of irritability as mechanical, though it needs to be said that his basic attitude on the issue cannot be characterized as strictly reductionist. With its emphasis upon external determination by means of tropic factors, Klebs' research programme greatly limited the appeal of Sachs' direction in morphogenetic research. This also became evident in cedidology, in which L. Diels (1913) finally emphasized one-sidedly the influence of the external environment and nutrition[14]. W. Magnus also spoke out decisively against the substance (chemical) theory of the origin of cecidia in his extensive monograph (1916). He held that cecidogenesis involves the mutual metabolic interaction of living cells (including parasitical cells), not the result of the activity of enzymes[15].

Here we touch upon the ambivalence of the concept of irritability : it could be understood and used both in a reductionist (see J. Loeb above) sense and in a non-reductionist sense. One of the most conspicuous reductionists of the time

12. See C. Herbst, *op. cit.*, compare also C. Herbst, *Formative Reize in der tierischen Ontogenese*, Leipzig, A. Georgi, 1901, VIII + 125 p.

13. W. Pfeffer, " Die Reizbarkeit der Pflanzen ", *Verhandlungen der Gesellschaft deutscher Naturforscher und Ärzte*, 65 (*Versammlung* 1893), 68-96.

14. L. Diels, " Der Formbildungsprozeß bei der Blütencecidie von Lonicera Untergatt. Periclymenum ", *Flora*, 105 (1913), 184-223.

15. W. Magnus, *Die Entstehung der Pflanzengallen, verursacht durch Hymenopteren*, Jena, G. Fischer, 1914, 160 p.

among plant physiologists was J.Ch. Bose, who attempted to show that living beings react to their surroundings in a fashion that is in principle the same as inorganic matter : " Thus living response in all its diverse manifestations is found to be only a repetition of responses seen in the inorganic "[16]. He tried to explain irritability as a sudden change in the molecular structure of living matter, manifesting themselves as movement : " The shock of stimulus causes molecular derangement in the tissue of the plant, and it is the fundamental molecular change that finds expression in mechanical movement. It finds independent expression also in electrical movement "[17]. In his numerous studies Bose always conceived the irritability of plants as a mechanical effect which might be recorded and measured by gadgets for the construction of which he was also famous[18].

In contrast to this, other plant physiologists attempted to find phenomena in their subjects such as those already known from conditions among higher organisms, i.e. animals. The results of this non-reductionist approach often produced unexpected discoveries even though they could not confirm their initial theoretical notions. Thus in Prague, B. Nemec looked for specific structures for inducing irritation in plant plasma (i.e. a formation similar to animal nerves ; his colleague at the German university there, F. Czapek, even considered that the existence of a reflex arc could be demonstrated in plants) and so discovered an earth-gravity perception mechanism using starch-containing grains in the cells[19]. Nemec was originally a zoologist and in formulating his statolith theory of geotropism he was able to base himself on his knowledge of the construction of gravity-sensing organs in crustaceans, which also served as a model. It should be pointed out that G. Haberlandt, the author of an essentially identical theory of the perception of earth's gravity in plants, who arrived at his explanation at the same time, was at that time engaged in a review and critique of Nemec's ideas on conductive plasmatic structures[20]. F. Noll, the botanist, also put forward a hypothetical model at that time of the sensory apparatus involved in perceiving earth's gravity and changes in the position of a plant, built on the model of the receptor already described in lower animals though he imagined it to be a very small formation, even submicroscopic[21].

16. J.Ch. Bose, *Response in the Living and Non-living,* London ; New York ; Bombay, Longmans, Green & Co., 1902, 189.

17. J.Ch. Bose, *Plant Response As a Means of Physiological Investigation,* London ; New York ; Bombay, Longmans, Green & Co., 1906, 11.

18. See, for example, J.Ch. Bose, *Collected physical papers,* London, etc., Longmans, Green & Co. 1927, XIII + 404 p.

19. B. Nemec, *Die Reizleitung und die reizleitenden Strukturen bei den Pflanzen,* Jena, G. Fischer, 1901 ; F. Czapek, " Weitere Beiträge zur Kenntnis der geotropischen Reizbewegungen ", *Jahrbücher für wissenschaftliche Botanik,* 32 (1898), 175-308.

20. G. Haberlandt, " Über Reizleitung im Pflanzenreich ", *Biologisches Centralblatt,* 21 (1901), 369-383.

21. F. Noll, *Über heterogene Induktion,* Leipzig, 1892, 40 ; " Über Geotropismus ", *Jahrbücher für wissenschaftliche Botanik,* 34 (1900), 457-506.

Only Nemec and Haberlandt managed to " see " such an apparatus in a normal plant cell, understandably in certain tissues only (Nemec particularly studied the rootcap, Haberlandt studied starch sheaths in stems)[22].

Even more important results than the study of geotropism were achieved for the physiology of irritability by experiments for inducing phototropic stimulus. An important discovery was made by the Danish botanist P. Boysen-Jensen, who in 1910 published a proof of the inducement of such an irritation around the rear side of a coleoptile in an oat caryopsis. The author attempted to interrupt the inducement by inserting slate chips into the presumed paths ; he impaired them by cutting and so forth. Moreover he found that the stimulus is safely conveyed across a blocked space by means of a little bridge made out of gelatine. Cuts on their own did not suffice to prevent contact and inducement : other experimental results came about in a moist environment and others, by contrast, in a dry place or under water[23].

He carried on in many respects from his predecessor, Fitting, who was working with cross-sections (1907) and from Rothert, who as early as 1896 had found that a break in a vascular bundle in a coleoptile does not cause cessation of the tropic reaction. The author completed his experiments at the Pfeffer Institute in Leipzig which also made it clear in which conceptual framework his experiments were being performed. It was there that the Hungarian botanist, A. Páal, also confirmed and supplemented his results (1914, 1918)[24]. The basis of irritation inducement is the diffusion of the active substance dissolved in water. This substance comes from the tip of the coleoptile and is a growth regulation agent. It is worth mentioning that as early as 1897, F. Czapek had proposed recognition of a chemical substance as an agent in the inducement of tropic excitement — though he had mistakenly identified it as homogentisic acid[25]. Although himself a prominent biochemist, Czapek had always rather tended towards explanations of tropisms in terms of physics.

Boysen-Jensen and Páal were united in refuting this physics-oriented solution to the stimulus inducement issue put forward by Fitting in 1907 on the strength of his still rather rudimentary experimental methods (Fitting explained

22. B. Nemec, " Über die Art der Wahrnehmung des Schwenkkraftreizes bei den Pflanzen ", *Berichte der Deutschen Botanischen Gesellschaft,* 18 (1900), 241-245 ; G. Haberlandt, " Über die Perception des geotropischen Reizes ", *Ibid.,* 18 (1900), 261-272. See also J. Janko, " Prazsky prinos k objasneni geotropismu u rostlin : Friedrich Czapek a Bohumil Nemec " (Prague contribution to explanation of geotropism in plants), *Dejiny ved a techniky,* 11 (1978), 29-45.

23. P. Boysen-Jensen, " Über die Leitung des phototropischen Reizes in Avenakeimpflanzen ", *Berichte der Deutschen Botanischen Gesellschaft,* 28 (1910), 118-120.

24. A. Páal, " Über phototropische Reizleitungen ", *Berichte der Deutschen Botanischen Gesellschaft,* 32 (1914), 499-502, " Über phototropische Reizleitung ", *Jahrbücher für wissenschaftliche Botanik,* 58 (1918), 406-458.

25. F. Czapek, " Über einen Befund an geotropisch gereizten Wurzeln ", *Berichte der Deutschen Botanischen Gesellschaft,* 15 (1897), 516-520.

the inducement of excitement even across cuts by means of polarization)[26]. Boysen-Jensen later had to protest against the pioneers in the theory of growth hormones, particularly Went, who understandably no longer used the term " irritation ", or " stimulus " and who was able to explain everything in terms of chemistry. The Danish botanist correctly pointed out that the hormonal theory of growth itself would have been unthinkable without the application of physiology of irritability ideas and methods (1935). He argued for saving the term " stimulus " (*Reiz*), combining it with the term *Auslösung : Ein Reiz wirkt durch Auslösung ; der Erfolg ist daher sowohl durch den äußeren Reiz als auch durch die innere Organisation der Pflanze oder des Pflanenorgans bedingt*[27]. Boysen-Jensen declared that the term stimulus was defined better by Pfeffer than by Sachs. In connection with the discovery of the transfer of irritation by a chemical substance, there were also certain hopes involving the possibility of applying Sachs' organogenic substance research programme conception. However, in the course of very intensive research into conductive substances, which also had aspects that were of practical use (influencing growth) it was shown that non-specific substances were involved, with collective results for the most varied kinds of plants and their organs. Research into growth hormones (auxins and related substances) soon completely overshadowed the traditional subject of the physiology of irritability. In 1932 Laibach even proved that Fitting's *Reizstoff* from orchid pollen and growth hormones are one and the same chemical compound. An elucidation of the chemical base of auxins was in fact the fruit of intensive research by F. Koegl and his colleagues in the 1930s[28].

F.W. Went at the end of the 20s and the beginning of the 30s elaborated a theoretical conception for the hormonal theory of growth which he believed could eliminate the now redundant remains of the conceptual and methodological framework around classical physiology of irritability. He designated the active substance that regulated growth as growth-substance (*Wuchsstoff*) which in fact did not act qualitatively as Sachs and his followers had presumed, but entirely quantitatively. Went explained tropism as the consequence of changes in the formation and direction of the transport of growth substance[29]. Special (phototropic, geotropic) *Reizstoffe*, anticipated by earlier researchers, did not in

26. H. Fitting, " Die Leitung tropistischer Reize in parallelotropen Pflanzenteilen ", *Jahrbücher für wissenschaftliche Botanik,* 44 (1907), 177-253 ; " Lichtrezeption und phototropische Empfindlichkeit, zugleich ein Beitrag zur Lehre vom Etiolement ", *Ibid.*, 45 (1907), 83-136. Fitting emphasized the directe influence of the sunlight on the plant tissues : " Für die Lichtform des Internodium als wesentliche Faktoren der Gestaltung nur eine direkte Lichtreizung des Hypokotyls und eine direkte Lichtreizung der Koleoptile mit Transmission des Lichtreizes auf das Hypokotyle von Bedeutung sind ", *Ibid.*, 127.

27. P. Boysen-Jensen, *Die Wuchsstofftheorie und ihre Bedeutung für die Analyse des Wachstums und der Wachstumsbewegungen der Pflanzen,* Jena, G. Fischer, 1935, 152.

28. Compare P. Boysen-Jensen, *op. cit.*, 12 etc.

29. F.W. Went, " Wuchsstoff und Wachstum ", *Recueil des travaux botaniques néerlandais, 25* (1929), 1-116.

fact exist : *Es hat sich also herausgestellt, daß in allen näher untersuchten Fällen eine direkte quantitative Beziehung zwischen Auxin und angeregtem Wachstum besteht, und es gibt überhaupt keine Gründe mehr, die prinzipielle Ungleichheit der Wachstumsbeeinflussung anzunehmen, also die Wachstumsanregung als Reizerscheinung aufzufassen*[30]. Similar conclusions were made by the Russian physiologist N.G. Kholodny. From the 1930s their jointly developed theory of growth substances became the leading research programme in plant physiology and entirely overshadowed Pfeffer's classical physiology of irritability. This does not mean that different concepts and approaches did not continue to be applied. The hypotheses of B. Nemec on organogenes and M.Kh. Tschailakhyan's experiments at the same time again awoke initial guarded interest in Sachs' conception of formative substances[31].

In conclusion it can be said that the mutual inspiration on either side of the boundary between developmental mechanics and the physiology of irritability in experimental botany led to results of great theoretical and practical importance. Both leading conceptions had the advantage of allowing for both a reductionist and a vitalistic (or at least a non-reductionist) interpretation of results and especially posed questions which could be resolved both by observation and by experimentation. A certain dogmatism and rigidity on the part of some research programmes understandably made it difficult for more fruitful work to be done (this surprisingly was the case for example with Klebs' conception). Despite the most varied reservations and criticisms, the physiology of irritability actually represented a research programme with a great supporting capacity, which was a necessary condition for further penetration into the molecular bases of phenomena studied by experimental botany.

ACKNOWLEDGEMENTS

The author wishes to thank the *Deutsche Forschungsgemeinschaft, Deutscher Akademischer Austauschdienst* for supporting grants (in the years 1989, 1991, and 1995 respectively).

30. F.W. Went, " Allgemeine Betrachtungen über das Auxin-Problem ", *Biologisches Centralblatt*, 56 (1936), 449-463 (quot. p. 455). Compare his sentence (*Ibid.*, 453) that *eine Auffassung der Hormone als Reizstoffe, als Auslöser physiologischer Spannkräfte, welche ganz allgemein Prozesse in Gang setzen, keine Erklärung der quantitativ abgestuften Korrelationen gibt.*

31. On the fate of Sachs' concepts in the development of modern plant physiology see A. Pirson, " Julius Sachs : Arbeit und Denken aus der Sicht der neueren Pflanzenphysiologie ", in H. Gimmler (ed.), *Julius Sachs und Pflanzenphysiologie heute*, Würzburg, Verlag der Physik.-Med. Gesellschaft 1984, 115-162, and W. Hartung, " Der Beitrag von Julius Sachs zur Entdeckung der Phytohormone ", *Ibid.*, 167-180.

EXPLANTATION IN EARLY 20th CENTURY CYTOLOGY : FROM MECHANISTIC TO ORGANISMIC CONCEPTS

Brigitte HOPPE

THE BEGINNING OF EXPERIMENTS WITH ISOLATED TISSUES AND CELLS BASED ON A MECHANISTIC EPISTEMOLOGY

In many biological and medical areas the methods working with isolated tissues and cells are practised daily in our time. The beginning of this practice in German and other European countries is not well known. Only a few data concerning the earliest experimenters were mentioned in the historiography[1]. Around the middle of the 19th century mainly the blood cells, micro-organisms and some early embryonic stages were the observed isolated single animal cells, but their cultivation remained difficult until the last quarter of the 19th century. At the same time it was only possible to isolate tissues by the help of fixation methods for histological preparations, that means to observe dead tissue parts under the microscope. At first in the area of developmental physiology zoologists, namely the Frenchman L. Chabry and the German Wilhelm Roux tried to make experiments with isolated embryonic animal tissues in the eighties[2]. But the cultivation problem remained still unresolved. It was not before the year 1907 that Ross Granville Harrison at Yale University observed the growth of isolated nerve cells in vitro. Then, the French surgeon Alexis

---

1. Mainly R. Harrison's and A. Carrel's findings are mentioned in the following books : A. Hughes, *A History of Cytology*, London, New York, Abelard-Schuman, 1959, 121 f., 137 ; P. Diepgen, *Geschichte der Medizin*, vol. 2, pt. 2, 2nd ed., Berlin, W. de Gruyter, 1965, 54, 113, 292 f. ; J.M. Oppenheimer, " Ross Harrison's Contributions to Experimental Embryology ", *Essays in the History of Embryology and Biology*, by J.M. Oppenheimer, Cambridge, Mass., London, M.I.T. Press, 1967, 92-116. ; *Geschichte der Biologie*, ed. by I. Jahn and others, 2nd ed., Jena, G. Fischer, 1985, 488, 495, 675. ; M. Borell, *Album of Science : The Biological Sciences in the Twentieth Century*, New York, Charles Scribner's Sons, 1989, p. 64 f. Figs. 90 and 91, p. 259 Fig. 357.

2. F.B. Churchill, " Chabry, Roux and the Experimental Method in Nineteenth Century Embryology ", *Foundations of Scientific Method : The Nineteenth Century*, ed. by R.N. Giere and R.S. Westfall, Bloomington, London, Indiana University Press, 1973, 161-205.

Carrel working in New York since 1906 could improve the methods of cell cultivation[3]. From that time on, starting from the United States, physicians and biologists began to use the methods of tissue culture and of the so-called " explantation " (notion created by Wilhelm Roux in 1905) of cells more and more in different biological research fields as well as in medical pathology.

But on the details of the introduction of the new methods into different areas of biological and medical research, especially in European countries, the historical literature did not publish much information. Also the historic conditions, the consequences of the development of the new methods and the significance of those scientific transformations were not a topic of a historic analysis. Therefore, I would like to discuss some unknown events concerning those topics and their historic importance in the following contribution focusing mainly on the details of the historic development in Germany and Europe. I studied the main original publications of scientists involved in researches of the new field and added some manuscript documents coming from main figures in that history. Moreover, I analyzed the first international journal for experimental researches with isolated cells. Although this first special journal for the new area was founded in Germany, it started with an international editorial board and published many papers from foreign authors. Therefore, my contribution will discuss also some historic aspects of the international biology of the first half of our century important for the history of biology in general.

For the earliest period using methods of transplantation or explantation mainly in embryonic experiments, we have to underline that their epistemological fundamentals were characterized by the strong reductionistic and especially mechanistic thinking in the biology of the late 19th century — mainly in Germany. One of the first experimenters introducing and propagating experiments with isolated embryonic tissues since 1885 was Wilhelm Roux, the founder of the so-called *Entwickelungsmechanik*. He defined the methodological aim of this experimental biological field in the following manner :

" The *Entwickelungsmechanik* follows the aim to reduce the organic developmental processes (*Gestaltungsvorgänge*) finally to the smallest number and to the most simple modes of efficacy (*Wirkungsweisen*) and to find out the intensity of their efficacy (*Wirkungsgrößen*) ; moreover, in this research area the processes of transformation of matter and of force [he meant energy] (*Stoff- und Kraftwechsel*) causing the effects shall be research subjects "[4].

Many experimenters, e.g. the collaborators of Roux and several pathologists, followed the same interests in the years around 1900. They know how to cultivate tissue fragments in a medium with blood plasma, eggs, egg fragments and embryos of lower animals in albuminous solutions. All these media were

3. See in the literature listed in fn. 1.
4. W. Roux, *Die Entwickelungsmechanik - ein neuer Zweig der biologischen Wissenschaft*, Leipzig, Wilhelm Engelmann, 1905, 19.

isolated from organisms and became lifeless ; often they were placed on small glass plates, in order to watch them directly under the microscope. But they remained fresh only for a short time, what was caused also by failing of aseptic methods. Mainly the experimenters treated the preparations in a relatively rough manner with inorganic chemicals and by mechanic irritations like extension or pressure (Roux)[5], looking for the effects of the treatment, which were e.g. incomplete cell division or excessive cell growth. Remarkable from the viewpoint of early experimentation with cells were the systematic experiments performed by the brothers Hertwig, carried out between 1885 and 1892. In order to gain insight into the phenomena and conditions of fertilization and the first stages of cell division and embryo development — discussing also some observations described by W. Roux — they treated sea urgin eggs with chemical agents, heat and shaking. They applied for instance nicotine, strychnine, morphine, quinine, chloral and others and observed different malformations in the fertilization process and, if successful, in cleavage, *Karyokinesis* (mitosis) and further stages of cell division. Concerning their interpretation of the results it is to notice, as Paul Weindling pointed out in 1991, that they favoured a " causal analytical approach " (*eine mechanische Erklärung der Lebens-vorgänge*), but must admit the difficulties of a clear statement in such a " dark area " of biology. O. and R. Hertwig pointed out in 1887 that they were searching for a " mechanical explanation ", although the life phenomena seemed to become more complex during the experiments. Therefore, they wanted to resolve the complex life processes into their components at first[6]. Then in 1897 Oscar Hertwig began to criticize the extreme mechanistic interpretations, especially those presented by W. Roux[7]. Soon other biologists began to improve the methods and enlarged the questions during the first decade of our century.

But because of the deaths of several scientists of the first generation around the 1920s and the strong restraining of science in general following the first World War, the development was hampered. During the same period until the twenties more than 70 authors of foreign countries, mainly the USA, England, France, Italy, Russia, and Japan published on their researches with the methods of tissue culture. Then only after two or three years, we can state more and more German authors publishing on the area during the twenties and thirties ;

5. *Idem*, " Meine entwicklungsmechanische Methodik ", *Handbuch der biologischen Arbeits-methoden*, ed. by E. Abderhalden, Abt. V, Teil 3, A, Heft 4, Berlin, Wien, Urban & Schwarzen-berg, 1923, 539-616, see 587-591.

6. O. und R. Hertwig, " Experimentelle Untersuchungen über die Bedingungen der Bastard-befruchtungen ", *Jenaische Zeitschrift für Naturwissenschaften*, 19 (1886), 121-165; *Idem, Über den Befruchtungs- und Teilungsvorgang des tierischen Eies unter dem Einfluß äußerer Agentien* (= Untersuchungen zur Morphologie und Physiologie der Zelle, Heft 5), Jena, G. Fischer, 1887, 147 ; See also P.J. Weindling, *Darwinism and Social Darwinism in Imperial Germany : The Contribution of the Cell Biologist Oscar Hertwig (1849-1922)* (= Forschungen zur neueren Medizin- und Biologiegeschichte, Bd. 3), Stuttgart, New York, G. Fischer, 1991, 115-118.

7. O. Hertwig, *Mechanik und Biologie* (= Zeit- und Streitfragen der Biologie, H. 2), Jena, G. Fischer, 1897, 98-196.

that means we see rising experimental cell research during the so-called Weimar Republic in Germany, although the post-war and then the international economic crisis did not afford suitable conditions to further scientists and scientific institutions. These statements provoke some explanation and a deeper historic analysis of the activities and of the scientific significance of the researches in the biomedical field during that period in Germany. In the following I would like to present some results of my studies of original publications, some manuscript sources and especially the contents of the first international special journal for experimental cell research founded in 1925 in Germany. For the first period of flourishing of the new area in Germany we may discern several stages of the introduction of the new methods and of their spreading through many biological disciplines.

I. 1911-1919 : Testing of methods by individual biologists

Mainly scholars in the field of developmental physiology were using and testing methods of transplantation and then of explantation : the physician and zoologist, assistant of Roux and later professor in Halle a. S. Albert Oppel (1863-1916) since 1911[8], Hirsz-Elia Ossowski since 1912, while working also with Roux in Halle a. S., and the zoologist and embryologist in Würzburg and Heidelberg Hermann Braus (1868-1925)[9]. At first, they were testing the methods watching growth or development of different tissues in vitro, observing also movement of epithelial and embryonic cells. They learned the methods to cultivate living embryonic cells on small glass plates under the microscope from publications and demonstrations by Ross Granville Harrison around 1910 (H. Braus, 1911), who was a well known personality in Germany, since he had absolved his medical studies at the university of Bonn 1892-1893, then in 1895-1896, and had visited Germany several times, giving also lectures at different places and publishing in the German language[10].

With much more intensity two young German biologists learned the new methods by direct training in the laboratorium of Harrison : Rhoda Erdmann and Richard Goldschmidt. One of them became the real promotor of the new area in Germany since the twenties. Both, coming from the Zoological Institute of the University of Munich under Richard Hertwig, had made studies in developmental physiology and cytology. Erdmann had worked with protozoa and Goldschmidt with insects (the butterflies Lymantria). Both had worked as guests with Harrison at Yale University, Erdmann from 1913-1914 and Goldschmidt from November 1914 until springtime 1915. Both published their first results in the area in American journals.

8. A. Oppel, Gewebekulturen und Gewebepflege im Explantat, Braunschweig, Vieweg, 1914.

9. H. Spemann, " Hermann Braus + ", Die Naturwissenschaften, 13 (1925), 253-261.

10. J.M. Oppenheimer, " Ross Granville Harrison ", Dictionary of Scientific Biography, 6, New York, Charles Scribner's Sons, 1972, 131-135.

Goldschmidt could demonstrate that the formation of spermatozoa observed from early stages of cell division on is dependent on the state of development of the follicle cells (published 1915-1917). After his return from the USA in 1919 he was appointed head of the department for genetics at the Kaiser-Wilhelm-Institut (hereafter KWI) for biology in Berlin and didn't continue working especially with the methods of explantation and tissue culture[11]. Quite different was the scientific career of the female colleague.

Rhoda Erdmann[12] analyzed the life-cycle of protozoa (*Trypanosoma brucei*). Then she demonstrated the usefulness of explantation methods in immunological research. She could diminish the efficacy of highly infectious serum of the chicken pestilence by the help of bone-marrow cultivation (1913), than it became possible to immunise chickens actively with the manipulated material (1920). After coming back to Germany in 1919 she tried to introduce the methods of isolated cell cultivation in a special research institution. In spite of many difficulties she became a pioneer by practising the methods in biomedical research in Germany.

II. 1919-1925 : Institutionalization of the new area in Germany

During the second period from 1919 to 1925 we can state a remarkable intensification of research and the institutionalization of the area by several new foundations of scientific activity in Germany. But the beginning was very hard in the first post-war years. When Rhoda Erdmann applied for a position in 1919, she got around fifty refusals in half a year. Also the biologists of the KWI in Berlin-Dahlem, including Richard Goldschmidt who himself was just returned, could not procure a position. But thanks to him and other colleagues working in the field of developmental physiology, namely Roux, Driesch[13], Curt Herbst and August Bier, she got references. Then, she could begin to work in a small, but own section specialized as *Abteilung für experimentelle Zellforschung* at the university institute for cancer research at the Charité in Berlin in 1919. Since in the same year German female academicians were admitted to gain the qualification of the habilitation at universities for the first time, she acquired this degree in the philosophical faculty in 1920 and also in the medical faculty in 1923. One year later she was appointed professor without an adequate salary. Not before the year 1929 she was appointed officially extra-ordinary professor and she became the head of the special department with the name " Universitätsinstitut für experimentelle Zellforschung " in 1930.

11. R.B. Goldschmidt, *Im Wandel das Bleibende*, Hamburg, Berlin, Paul Parey, 1963, 62-65, 80-84, 90-96, 178-180.

12. B. Hoppe, " Die Institutionalisierung der Zellforschung in Deutschland durch Rhoda Erdmann (1870-1935) ", *Biologie Heute*, Nr. 366 (= Beilage zur Zeitschrift *Naturwissenschaftliche Rundschau*, 42, H. 7) (1989), 2-4, 9.

13. R. Mocek, *Wilhelm Roux - Hans Driesch. Zur Geschichte der Entwicklungsphysiologie der Tiere* (= Biographien bedeutender Biologen, Bd. 1), Jena, G. Fischer, 1974.

Further activities were also important for the advancement of cell research. Rhoda Erdmann published research reviews concerning the years 1914 to 1920, continued for the following years[14]. She was teaching the cell and tissue cultivation methods to students of the biomedical sciences and published the first textbook of the area under the title *Praktikum der Gewebepflege oder Explantation, besonders der Gewebezüchtung*, the second edition of which appeared in 1930[15]. In 1922, the anatomist and embryologist Hermann Braus published the methodical introduction in the handbook of biological methods (*Handbuch der biologischen Arbeitsmethoden*) edited by Emil Abderhalden under the title *Methoden der Explantation (Gewebekulturen in vitro)*[16]. At the same time the Dane Albert Fischer (1891-1956) working temporarily in New York with Carrel, then in Copenhagen was writing a *Manual for the Biology of Isolated Tissue Cells* since 1920, which was published in 1925 in English and enlarged in German in 1927 and 1930[17]. The work of this researcher was also much appreciated by the biologists at the KWI in Berlin, so he was invited as visiting professor for one year in 1926. The interest in the area grew in Germany and in general. In the special bibliography for the years 1920 to 1925, compiled by R. Erdmann, we find 20 German authors and 25 excellent specialists of the most civilized nations from the USA through Russia to Japan. They dealt mainly with histological, cytological, embryological, physiological and pathological problems as well as with the improvement of the equipment and methods. From 1925 on research works in the area flourished at the main scientific places of the world.

III. 1925-1939 : Intense scientific and medical research on an international level

In Germany Berlin was now a center where the teams of R. Erdmann and A. Fischer experimented with animal cells. Their studies were completed by botanists : Gottlieb Haberlandt dealt with the physiology of cell division of

    14. R. Erdmann, " Einige grundlegende Ergebnisse der Gewebezüchtung aus den Jahren 1914-1920 ", *Ergebnisse der Anatomie und Entwicklungsgeschichte*, ed. by E. Kallius and others, = *Zeitschrift für die gesamte Anatomie*, Abt. 3, 23 (1921), 420-500 ; *Idem*, " Verzeichnis der in den Jahren 1920-1924 erschienenen Arbeiten aus dem Gebiete der Explantation ", *Archiv für experimentelle Zellforschung*, 1 (1925), 130-144 ; *Idem* und E. Pinzke, " Verzeichnis der in den Jahren 1925 und 1926 erschienenen Arbeiten aus dem Gebiete der experimentellen Zellenlehre, besonders Gewebezüchtung (Explantation) ", *Ibidem*, 4 (1927), 125-142.
    15. R. Erdmann, *Praktikum der Gewebepflege oder Explantation, besonders der Gewebezüchtung*, 1st ed. Berlin, Springer, 1922, 2nd ed. Berlin, Springer, 1930.
    16. H. Braus, " Methoden der Explantation (Gewebekulturen in vitro) ", *Handbuch der biologischen Arbeitsmethoden*, ed. by E. Abderhalden, Abt. V, Teil 3, A, Heft 3, Berlin, Wien, Urban & Schwarzenberg, 1923, 517-538.
    17. A. Fischer, *Tissue Culture* (= Kobenhavns Universitet, Inst. for Almindelig Pathologi = Contributions from the Univ. Inst. for General Pathology, Copenhagen, vol. 4, 1926) ; *Gewebezüchtung, Handbuch der Biologie der Gewebezellen in vitro*, German transl. by F. Demuth, 2nd ed., München, R. Müller & Steinicke, 1927 ; 3rd ed., *Ibid.*, 1930.

plants[18], whilst Ernst Küster in Gießen discussed experimentally the morphology, physiology, and pathology of plant cells since 1920[19]. Both botanists discussed their results also with the zoologists working on experimental cytology.
But still in the twenties the cytologists stated that their field was not appreciated enough as an area of special researches important for all biomedical sciences. Therefore, they founded a new special forum of communication for the
advancement of their science in 1925 : the first special international journal for
experimental cell research entitled *Archiv für experimentelle Zellforschung,
besonders Gewebezüchtung (Explantation)*. It appeared in Jena, Germany, published by the Gustav Fischer-publishers. It was founded by 15 biologists from
different nations : six Americans, among them Harrison and Leo Loeb, two
Russians, an Englishman, a Dane, an Italian, a Frenchman and a Swiss. The
editorial board included the German botanist Ernst Küster and Rhoda Erdmann, who performed the main work as editor until her death in 1935. Moreover, 38 authors from other nations showed by their contributions the world-
wide diffusion of the field. That archive was an international organ until 1938.
That year appeared volume 22 which included the proceedings of the fifth
international congress on cytology in Zürich, Switzerland, which was visited
by many biologists of Western countries, but also by participants from Russia
(1), Poland (2) and Israel (1). The following volumes contained a majority of
German contributions, to which were added only articles by Italian, Hungarian,
Dutch and Scandinavian authors. Only in volume 23 (1939) three papers by
Americans were added. In volume 24 (1942) the editors mentioned some difficulties caused by the war. Finally, the last small issue number 25, appeared
in 1944.

In the first volume the editors characterized the research programme and the
main tendencies of their work in 1925 with the following words : " We intend
to show to the scientists working in biomedical sciences, that studies on living
cells themselves and working with isolated living cells are required by prefer-
ence for the next future. To esteem studies of the isolated living cell, of living
tissue, whether health or sick one, having the same value as observations with
fixed and stained preparations, that is the desirable aim of our didactics,
[…] "[20].

18. See E. Höxtermann, " Cellular 'elementary organisms' in vitro. The early vision of Gottlieb
Haberlandt and its realization ", *Physiologia Plantarum*, 100 (1997), 716-728.

19. E. Küster, *Die Pflanzenzelle*, Jena, G. Fischer, 1935 ; *Idem, Experimentelle Zellforschung*,
Jena, G. Fischer, 1948 ; 2ⁿᵈ ed. 1949.

20. R. Erdmann, " [Vorwort zum 1. Band, Februar 1925] ", *Archiv für experimentelle Zellforschung*, 1 (1925), [Vorwort].

MAIN RESULTS OF INTENSE INTERNATIONAL CELL RESEARCH IN THE MIRROR
OF THE NEW ARCHIVE BETWEEN 1925 AND 1940

Let us ask now : How effective were the efforts in the new area ? Could the
ideal purpose be realized ? May we state a development of the field in Germany and other European countries during the twenties and thirties ? I would
like to discuss these questions by analyzing the main results of cell research
published in the new archive between 1925 and ca. 1940[21]. The majority of
papers in the 25 volumes came from foreign laboratories. Concerning the German contributions we find repeatedly the half part of them produced by assistant researchers in the laboratory of Rhoda Erdmann in Berlin. In general, the
subjects of research were mainly taken from animal tissues, sometimes from
human or plant tissues, rarely from micro-organisms (such as bacteria and protozoa). In the first volumes some virological and immunological problems
were discussed, which almost disappeared after 1928. In many studies the biologists observed the behaviour of different tissues and cell types in vitro by
dealing with morphological and physiological questions, e.g. heart cells in very
early embryonic states showed the spontaneous rhythmic motion of their later
function. Furthermore, the processes of regeneration of epithelial cells and
phagocytosis could be observed in detail. In cell colonies' cultures the formation of immune particles and substances were studied. During the whole
period, special questions of the methods in general and staining methods in
particular as well as improvements of apparatus were discussed. To the empirical research studies were added reports and bibliographies.

It is to notice that papers on that field from which the explants cultivation
originated, the embryology, appeared rarely, since there existed several special
journals on it. But histological and cytological contributions based on the traditional area of microscopic anatomy filled often a quarter of the articles of one
volume. Haematological studies with blood cells were included in every volume. Pathological papers were dominated since 1928 by cancer research. But
on topics of some fields which are using the methods of tissue and cell culture
nowadays, was only a small number of articles published, namely on biochemical, physicochemical, pharmacological, toxicological and chemothera-peutical
problems. Then, contributions on radio biology were multiplied since 1931.
From 1931 on, questions on modern genetics were treated only by foreign
authors. By comparing the main themes of the articles we can state that the
methods of cell colonies' cultivation were introduced more and more into the
main areas of biomedical sciences including the then modern directions of
radio biology and genetics, though the last topics were restrained in the cytological archive.

21. R. Erdmann and others [editors], *Archiv für experimentelle Zellforschung, besonders
Gewebezüchtung (Explantation)*, 1-17, Jena, G. Fischer, 1925-1935 ; E. Haagen [editor], *Idem*, 18-
25, *Ibid.*, 1936-1944.

Especially the German contributions were showing the tendency to deal with the new basic problems of experimental cytology. They wanted to know how tissues and cells explanted behave in vitro, how to use them, in order to investigate cytological, physiological and pathological problems. Especially in the laboratory directed by Rhoda Erdmann they worked on cell-physiological and cell-pathological problems beginning to focus on cancer research since the twenties. These activities helped to introduce the methods into medical research in Germany. Moreover, they obtained new knowledge on different cell types, on unknown cellular structures, functions, their development as well as physiological and pathological changes under special conditions. Therefore, they promoted basic research in cytology by the help of then modern experimental methods.

GENERAL HISTORIC ASPECTS DEDUCIBLE FROM THE EARLY DEVELOPMENT
OF THE CULTIVATION OF ISOLATED CELLS

Looking on the socio-economic and institutional conditions of the introduction and spreading of the methods of tissue and cell culture between 1920 and 1940, we state very unfavourable and poor political and economic conditions in Germany and other European countries. On the other side we found out that several institutions with a relatively simple and cheap equipment were founded. From these statements we conclude that only the excellent personal qualities like a high scientific qualification appreciated by the scientific community, an assiduous initiative with clever argumentation, and a strong will for realizing the aims were the basis for the success of the leading personalities, that was in Germany mainly Rhoda Erdmann. After the period of flourishing, only the destroying impacts of the events of the so-called " Total War " on the whole scientific life and system broke the fruitful work lastingly for around ten years.

Concerning the internal history of biology we state that from the new methods of tissue and cell culture working with minimal quantities of living material resulted new, much more exact observations on the conditions and effects of development and growth of different cell types as subjects of experiments. They led to new insights into cell physiology and pathology and to the beginning of an understanding of the complexity of cell life. That was especially important during that period between the appreciation of a very rough, mechanistic model of cell structure, for instance supported by Otto Bütschli with his pattern of a unique " honeycomb " (*Wabenstruktur*) of the whole protoplast around 1900 (until 1920)[22], and the possibility to observe more morphological and biochemical details, i.e. cell organelles by the help of the electronic micro-

22. O. Bütschli, *Untersuchungen über mikroskopische Schäume und das Protoplasma*, Leipzig, W. Engelmann, 1892.

scope since the thirties. The experimental methods of cell colonies' cultivation rendered it possible understanding the cell as a complex and very dynamic system.

After having discovered more differentiated qualities of cells, it was no more possible to adhere to the traditional simplifying mechanistic cell concept in general. Therefore, cytologists founding their studies on explantation experiments with living tissues and cells favoured a more organismic cell concept already since the twenties. But their interpretations were also different from those of some vitalistic concepts in embryology like that held by Hans Driesch. And the advanced interpretations of the cytologists in the thirties differed also from the conservative organicism of the brothers Hertwig mentioned above. These perceived more unsolved problems in the " dark " new research area of experimental embryology and cytology than answers on their questions. Although the following generation of biologists could not elucidate the whole " darkness ", they discovered much more specific qualities and faculties of living cells and tissues by their improved experimentation observing life phenomena directly without destroying the fundamental structures. They appreciated them as specific and performed the experiments more carefully, in order to conserve life. Therefore, they hoped to gain step by step more insights into that complex, specific organic system called cell.

" IMAGINE A CUBE FILLED WITH BIOLOGICAL MATERIAL " : RECONCEPTUALIZING THE ORGANIC IN GERMAN BIOPHYSICS, 1918-1945

Richard H. BEYLER

In a 1931 *Festschrift* commemorating the tenth anniversary of the Institute for Physical Foundations of Medicine at the University of Frankfurt, the director, Friedrich Dessauer, described the conceptual basis of his work in radiation biology. He employed, he said " a conception which works with necessary physical simplifications. In the place of the quite complicated field of effect of, say, a human or animal body, I imagine, as a physicist, a cube filled with homogenous, single-phase biological material... "[1].

That austere conceptualization of the organic was the origin of the " hit " or " target theory ". The theory's aim was to use radiation — the multiplicity of recently discovered particles and rays — to probe the structure of the organic. Central to the theory was the relatively new realization that the absorption of energy from radiation was quantized. Radiation impinging on living tissue was hence conceptualized as particles hitting a target. Based on hypotheses as to what biophysical process constituted a " hit " and what abstract or real entity constituted a " target ", one could compare statistical models with experimentally observed results. The approach was similar to the " cross section " technique familiar to physicists[2].

For a network of biophysicists, this new theory about the interaction between radiation and organisms signalled a sweeping transformation of their field. In 1936, Boris Rajewsky, Dessauer's erstwhile student and successor as

---

1. F. Dessauer, " Problemstellung und Theorie ", in *Zehn Jahre Forschung auf dem physikalisch-medizinischen Grenzgebiet*, ed. Dessauer, Leipzig, G. Thieme, 1931, 176-201, on 179.

2. Overviews include N.W. Timoféeff-Ressovsky and K.G. Zimmer, *Das Trefferprinzip in der Biologie*, Leipzig, S. Hirzel, 1947 ; B. Rajewsky and M. Schön, eds., *Biophysik*, 2 vols., Naturforschung und Medizin in Deutschland (FIAT Review of German Science), vols. 21-22, Wiesbaden, Dieterich'sche Verlagsbuchhandlung, 1948-49 ; K. Sommermeyer, *Quantenphysik der Strahlenwirkung in Biologie und Medizin*, Leipzig, Geest & Portig, 1952.

director of the institute after Dessauer's compelled emigration in 1934[3], issued a declaration of independence for biophysics in a *Denkschrift* proposing the establishment of a Kaiser Wilhelm Institute. " The powerful forward development of physics and physical technology [and]... the blossoming of atomic physics [open] fundamentally new paths in natural scientific knowledge ", proclaimed the proposal[4]. A Kaiser Wilhelm Institute for Biophysics was indeed created in 1937.

Radiation biology was the keystone of this instauration of biophysics. But as Dessauer's and Rajewsky's comments suggest, their approach was marked from the outset by a tension inherent in conceptualizing the object of study as " homogeneous matter " or " living substance " — the tension between biological variability and what Dessauer, self-consciously speaking as physicist, called " necessary simplifications ". The simplifying impulse was hotly contested.

The Frankfurt institute was the epicenter of target theory in Germany, but largely independently of Dessauer, similar ideas were developed in England, France, America, and elsewhere. Fernand Holweck of the Institut du Radium in Paris, collaborating with Antoine Lacassagne and Marie Curie, advanced something he called " statistical ultramicroscopy "[5]. In England, J.A. Crowther, a protégé of Ernest Rutherford, did parallel work[6]. In America, pathbreaking target theory research was started by E.U. Condon and H.M. Terrill[7].

It was in Germany that the consequences drawn from target theory by some of its advocates were the most sweeping. There was at least one significant international commonalty, however : almost all work in target theory was done either by scientists trained as physicists who switched their domain of inquiry or by collaborative teams of biologists and physicists. Rarely did biologists by themselves take up the instrumental and conceptual apparatus of target theory.

Dessauer's institute exemplified this trend. Dessauer, the son of an industrialist family from Aschaffenburg, became a designer of x-ray apparatus. The company he founded, VEIFA (Vereinigte Elektroinstitute Frankfurt-Aschaffenburg), was a leading supplier of medical x-ray equipment. After VEIFA merged

3. Dessauer was forced into exile by the Nazis, partly because of his politics (about which more below), partly because of a partially Jewish ancestry.

4. " Denkschrift zur Errichtung eines Kaiser-Wilhelm-Instituts für Biophysik zu Frankfurt am Main " (22 Oct. 1936), Magistratsakten 6603/33, Stadtarchiv Frankfurt.

5. The first of many articles in the *Comptes rendus* were A. Lacassagne, " Action des rayons K de l'aluminium sur quelques microbes ", 186 (1928), 1316-1317 ; F. Holweck, " Essai d'interprétation énérgetique de l'action des rayons X de l'aluminium sur les microbes ", 186 (1928), 1318-1319 ; [M.] Curie, " Sur l'étude des courbes de probabilité relatives à l'action des rayons X sur les bacilles ", 188 (1929), 202-204.

6. J.A. Crowther, " Some Considerations Relative to the Action of X-Rays on Tissue Cells ", *Proceedings of the Royal Society of London*, ser. B, 96 (1924), 207-211 ; and numerous subsequent articles.

7. E.U. Condon and H.M. Terrill, " Quantum Phenomena in the Biological Action of X-Rays ", *Journal of Cancer Research*, 11 (1927), 324-333.

with another firm, Dessauer turned increasingly to research — as well as political involvement in the Catholic-based Center Party. (He was elected to the
German parliament in 1924.) In 1920, Dessauer was appointed an honorary
professor at the University of Frankfurt, and the following year an Institute for
the Physical Foundations of Medicine was created around Dessauer's professorship[8].

In 1922 Dessauer began his forays into the statistical analysis of radiation
biology. He was struck by the idea that the amount of energy in a damaging
dosage of radiation was no greater than that in a cup of hot tea. Obviously it
was not the quantity of energy but the process by which it was absorbed which
produced damaging effects. Dessauer applied to this problem the realization of
the quantized nature of radiation. He modelled the absorption of radiation
energy in organic tissues as proceeding in discrete " deposits " which caused
localized elevations of temperature, resulting in turn in the localized denaturing
of proteins. Assuming that these deposits occur randomly in time and space, he
introduced an " effectiveness coefficient " to relate the dosage to the observed
effects. Statistically, the number of biological units which remained unaffected
after bombardment would be related to the dosage by an exponentially
decreasing function[9].

Dessauer's collaborators, Marietta Blau and Kamillo Altenburger, extended
the analysis to cases in which there must be more than one " hit " per object to
produce the effect. For a " one-hit " process this gave the original exponential
curve. For processes in which more than one " hit " was necessary, the dose-
response curve was sigmoid, with its specific shape dependent on the number
of hits required. Comparison of the experimentally observed dosage-response
curve with one of these theoretically generated curves would indicate how
many hits were entailed for the biological process in question to occur[10].

As noted earlier, Dessauer offered a hypothesis of what constituted a " hit ",
but that specific hypothesis was irrelevant to the more abstract character of his
approach. The effectiveness coefficient was for the time being a formalism to
make the experimental data fit the theoretical curve. Soon, however, a search
was underway to associating this coefficient with a real region or structure
within the organism. In other words, the effectiveness coefficient became conceptualized as a " target " ; an event of the required kind, such as an ionization,
taking place within the target constituted a " hit ".

8. On Dessauer's life and career, see his *Lebenslauf* in Dessauer Papers, Kommission für
Zeitgeschichte, Bonn ; W. Pohlit, " Friedrich Dessauer 1881-1963 ", website, http://www.phy-
sik.uni-frankfurt. de/paf/paf84.html, accessed 4 Dec. 1996.

9. F. Dessauer, " Über einige Wirkungen von Strahlen I ", *Zeitschrift für Physik*, 12 (1922), 38-
47.

10. M. Blau and K. Altenburger, " Über einige Wirkungen von Strahlen, II ", *Zs. Phys.*, 12
(1922), 315-320.

Thus in a 1924 paper Crowther developed a theory for the effect of x-ray irradiation on mitosis in protozoa which was formally similar to the Dessauer-Blau-Altenburger theory, but which postulated the physical presence of a sensitive target region instead. He also suggested an alternative concept of what constituted a " hit " : the production of an ion pair[11]. This latter hypothesis soon superseded the point-heat hypothesis. Holweck's statistical ultramicroscopy, as the name suggests, claimed to give information about the nature of real biological structures which were smaller than those visible in ordinary microscopes[12]. Also in Germany researchers quickly elaborated the Dessauer-Blau-Altenburger theory. One example was Richard Glocker, director of the x-ray laboratory at the Technische Hochschule Stuttgart. For his experiments on bean seedlings, Glocker introduced formulas based on the hypothesis that each of several targets in each biological object must receive n hits for the effect — here, stunted growth — to occur[13].

For its advocates, then, the new physical instruments of x- and other rays, plus the theoretical apparatus of target theory, promised access to a new level of biological inquiry. But there were also new assumptions — new to the life sciences, that is — about the nature of those biological objects. Above all, the target theory presupposed uniformity, as was standard procedure in the physical sciences. In the words of a biophysics text, " We... proceed from the presupposition, that all the objects which are subjected to the statistics behave completely similarly with respect to the radiation... This assumption is quite familiar to the physicist. Conversely, the biologist takes it as self-evident that biological objects always exhibit certain differences, a variability... "[14] . The targets, in other words, all had to act alike with respect if the theory was to work, but this was a notion heretofore not at home in the life sciences.

Indeed, the assumption of uniformity, in the eyes of many life scientists, begged the most significant question of biological inquiry. This complaint was voiced, for example, in a caustic critique of Dessauer's theory levelled by Hermann in the mid-1920s. Rather than assuming that organisms would act the same under similar circumstances, Holthused argued, it was the task of the life scientist to account for the differences in reactions and behavior. Hence the target theorists were ignoring what was for him the really interesting phenomenon : variations among individuals in their response to radiation. Holthusen

11. J.A. Crowther, *op. cit.* (see fn. 6).

12. F. Holweck, *op. cit.* (see fn. 5).

13. R. Glocker, " Die Wirkung der Röntgenstrahlen auf die Zelle als physikalisches Problem ", *Strahlentherapie,* 33 (1929), 199-205, and a series of subsequent articles.

14. K. Sommermeyer, *op. cit.* (see fn. 2), 46.

carried out his own experiments on seedlings, the results of which, he maintained, could only be explained by variable susceptibility[15].

Holthusen's objections were closely linked to his sensitivity to the potential medical implications of the theory. Was the conceptualization of the organic in such a way as to neglect variability the first step on the slippery slope to denying individuality in human beings ? A number of physicians and biologists attacked Dessauer on exactly this point which, in turn, had particular urgency due to the concerns raised by the " rationalization " and " modernization " in the Weimar era[16]. In terms developed by Jonathan Harwood in his recent study of the German genetics community, Dessauer was on the side of the modernizing pragmatists — and fits Harwood's pattern in terms of his non-traditional educational background and his high degree of political engagement, as well[17]. But Dessauer does present a difficulty for an unequivocal application of Harwood's model, since he was also intensely concerned with the broader philosophical implications of technology and science, seeking, within the framework of Catholic social political thought, a reconciliation between technical rationality and human values.

The austere simplifications at the basis of target theory were its source of power, but also its most vulnerable point. The approach could be elaborated and defended by taking more complex sets of factors into consideration. Starting with Dessauer's one-hit, one-target model, Blau and Altenburger had added multiple hits per event ; Glocker had added multiple targets per organism. Subsequent elaborations included temporal effects (" recovery " from hits), targets of varying size, and the like. More generally, individual variation could be subsumed under concept that a hit (e.g., ionization) in a target region only produced a certain probability that the observed biological effect would take place. From the target-theoretical point of view, all biological objects still behaved exactly the same with respect to a particular law, but this law was probabilistic in nature. Holthusen's critique still applied, however : was this not an unnecessarily abstruse way of conceptualizing the familiar notion of individual variability ?

Proponents responded with increasingly complex versions of the theory. By taking more factors affecting the response of the supposed targets into account, more refined statistical dosage-response curves could be created, which could be matched with more precision to experimental results. Further impetus came

---

15. H. Holthusen, " Über die Dessauersche Punktwärmehypothese ", *Strahlentherapie,* 19 (1925), 285-306, esp. 402 ; see also F. Pordes, " Zum biologischen Wirkungsmechanismus der Röntgenstrahlen ", *Strahlentherapie,* 19 (1925), 307-324.

16. For attacks on Dessauer, see A. Hessenbruch, " Processing Bodies : Radiotheraphy in the Weimar Republic ", unpublished MS, 1993 ; on modernization in the Weimar Republic see D.J.K. Peukert, *The Weimar Republic : The Crisis of Classical Modernity,* New York, Hill and Wang, 1993.

17. J. Harwood, *Styles of Scientific Thought : The German Genetics Community, 1900-1933,* Chicago, Univ. of Chicago Press, 1993.

with the introduction of new phenomena to investigate. Chief among these was the discovery in 1927, by H.J. Muller, that x-ray irradiation dramatically increased the rate of mutation in fruit flies. The burgeoning field of genetics was thereby opened to target-theoretical inquiry.

The application of target theory to genetics was cultivated assiduously at N.W. Timoféeff-Ressovsky's Department of Genetics of the Kaiser Wilhelm Institute for Brain Research in Berlin. Muller made an extended visit there on his way from the United States to the Soviet Union in 1933 ; about this same time an informal circle of biophysicists formed in Berlin, including physical experimentalist Karl G. Zimmer and physical theorist Max Delbrück[18].

The most influential product of this collaboration between genetics and physics in Berlin was a lengthy 1935 paper by Timoféeff, Zimmer, and Delbrück, " On the Nature of Genetic Mutation and Genetic Structure ". Applying a target-theoretical analysis to Timoféeff's extensive data, the authors concluded that x-ray mutagenesis was a " one-hit " process : namely, an ionization within a sensitive region of the order of magnitude of a large organic molecule. Buttressed by a thermodynamic argument from Delbrück, the authors concluded that the gene was, if not a large organic molecule, in any event a " group of atoms " of approximately that size. This paper was received as the clearest experimental evidence to date of the molecular nature of the gene[19].

Muller himself was sceptical of the target-theoretical line of reasoning, although the result confirmed ideas which he advanced for some time already, and although Muller was generally enthusiastic about the application of physics to biology. At a 1936 conference, Muller cautioned Timoféeff that the analysis was based on simplifying premises which were far from proven : the notion that there was a one-to-one correlation between ionizations in the sensitive region and mutation, the notion that there was no significant spatial transfer of energy in the mutation process, and so on[20]. Subsequently, Timofé-

18. M. Delbrück to N. Bohr, 30 Nov. 1934, in Bohr Scientific Correspondence, Niels Bohr Archive, Copenhagen ; 36-37 ; M. Delbrück, " A Physicist's Renewed Look at Biology : Twenty Years Later ", *Science*, 168 (1970), 1312-1315, on 1312 ; E.A. Carlson, *Genes, Radiation, and Society : The Life and Work of H.J. Muller*, Ithaca, Cornell Univ. Press, 1981, 188 ; L. Kay, " Conceptual Models and Analytical Tools : The Biology of Physicist Max Delbrück ", *Rivista di storia della scienza*, 2 (1985), 487-510.

19. N.W. Timoféeff-Ressovsky, K.G. Zimmer and M. Delbrück, " Über die Natur der Genmutation und der Genstruktur ", *Nachrichetn von der Gesellschaft der Wissenschaften zu Göttingen, Mathematisch-physicalische Klasse, Fachgruppe VI*, new ser., 1 (1935), 189-245. On the paper's reception, see inter alia R.C. Olby, *The Path to the Double Helix*, Seattle, Univ. of Washington Press, 1974, 232-234 ; M.F. Perutz, " Physics and the Riddle of Life ", *Nature*, 326 (1987), 555-558, on 556-557 ; D.B. Paul and C.B. Krimbas, " Nicolai V. Timoféeff-Ressovsky ", *Scientific American*, Feb. 1992, 86-92, on 88.

20. H.J. Muller, " Physics in the Attack on Fundamental Problems of Genetics ", *Scientific Monthly*, 44 (1937), 210-214 ; *idem*, " An Analysis of the Process of Structural Change in Chromosomes of *Drosophila* ", *Journal of Genetics*, 40 (1940), 1-66 ; *idem*, " Gene Mutations Caused by Radiation ", in *Symposium on Radiobiology : The Basic Aspects of Radiation Effects on Living Systems*, ed. J.J. Nickson, New York, John Wiley & Sons, 1952, 296-332, on 315.

eff and Delbrück said that the 1935 result only established a minimum size for the gene : the target region was not necessarily identical with the gene itself[21].

Zimmer's caution also increased ; he was increasingly alert to the relevance of Holthusen's earlier criticisms. In a 1941 survey, Zimmer conceded that the assumption of homogeneity was problematic, but argued that this did not invalidate the whole approach. Rather, factors of physical probability could be introduced so as to formally account for biological variability[22]. In some contexts, however, Zimmer conceded that observed dosage-response curves could best be explained by an assumption of biological variability. This was the case in an exchange with immunologist Richard Prigge over a target-theoretical interpretation of the process of immunization. After Zimmer's own statistical analysis of Prigge's experiments on tetanus immunization showed that a Gaussian distribution of susceptibility provided the closest match, Prigge politely but triumphantly wrote to Zimmer that " the introduction of target-theoretical representations for the explanation of my observations appears less urgent... "[23].

Delbrück, in his later recollections, maintained that the 1935 paper had been practically buried and forgotten[24], but Timoféeff and Zimmer continued to cite it prominently, though with increased caution, in numerous articles throughout the late 30s and into the 40s. Nor did target-theoretical inquiry take place on the periphery of German biology. As already noted, the Frankfurt institute under Rajewsky's directorship became a Kaiser Wilhelm Institute in 1937. And as Ute Deichmann's research has shown, Rajewsky, Timoféeff, and Zimmer were all in the top seven in support provided by the German Research Council for zoological research in this period[25].

Target theory was also attractive to quantum theorist Pascual Jordan who, in the late 1930s and into the '40s, saw in it the basis of a " quantum biology " which would uncover the " directing centers " of organic life. Jordan's vision of the importance of the technique went far beyond biophysics, however, in that he linked his quantum-physicalistic approach to biology with philosophical and political concerns : his attempts to delegitimize materialism and to

21. N.W. Timoféeff-Ressovsky and M. Delbrück, " Strahlengenetische Versuche über sicht bare Mutation und die Mutabilität einzelner Gene bei *Drosophila melanogaster* ", *Zeitschrift für induktive Abstammungs- und Vererbungslehre,* 71 (1936), 322-334, on 329, 331-332.

22. K.G. Zimmer, " Zur Berücksichtigung der 'biologischen Variabilität' bei der Treffertheorie der biologischen Strahlenwirkung ", *Biologisches Zentralblatt,* 61 (1941), 208-220.

23. K.G. Zimmer, " Zur Treffertheoretischen Analyse der Antigenwirkung ", *Naturwissenschaften,* 30 (1942), 452-453 ; R. Prigge and H.v. Schelling, " Zur Analyse der Antigenwirkung ", *Naturwissenschaften,* 30 (1942), 661 ; K.G. Zimmer to R. Prigge, 16 May 1942, and R. Prigge to K.G. Zimmer, 21 May 1942, in folder IIIA, Nachlass Prigge, Senckenbergische Bibliothek, Frankfurt.

24. L. Kay, *op. cit.* (see fn. 18), 221.

25. U. Deichmann, *Biologen unter Hitler : Vertreibung, Karrieren, Forschung,* Frankfurt, Campus, 1992, 81.

legitimize authoritarianism. In the context of the Third Reich, this was volatile stuff. Jordan's appropriation of target theory was yet another example of the way in which this reconceptualization of the organic could readily be made to carry cultural and social meanings[26].

Other leading German biophysicists, such as Timoféeff, Zimmer, and Rajewsky, though valuing Jordan's technical contributions, were less than warmly disposed towards his philosophical enthusiasms. For them, the institutionalization of biophysics was enough[27]. But by the mid-40s, the tension between the simplifying assumptions of target theory and the perpetual need to elaborate the theory to account for biological variability had come to a head. As the theory became more elaborate, its original elegant simplicity — the possibility of simply " reading off " the result from a plotted dosage-effect curve — was lost. The theory became more and more unwieldy. Researchers realized that simplifying assumptions of target theory could not account for the physiologically significant effects of energy transfers across (relatively) large distances and of complex biochemical processes[28]. With the rise of molecular biology, new re-conceptualizations of the physical foundations of the organic — simplifying in their own but quite different way — came into prominence.

26. M.N. Wise, " Pascual Jordan : Quantum Mechanics, Psychology, National Socialism ", in *Science, Technology, and National Socialism*, ed. M. Renneberg and M. Walker, Cambridge, Cambridge Univ. Press, 1994, 224-254 ; R.H. Beyler, " Targeting the Organism : The Scientific and Cultural Context of Pascual Jordan's Quantum Biology, 1932-1947 ", *Isis*, 87 (1996), 248-273.

27. See the introductions and conclusions to the texts cited in fn. 2.

28. E.g., H.J. Muller, " Gene Mutations ", *op. cit.* (see fn. 20) ; B. Rajewsky, " The Limits of the Target Theory of the Biological Action of Radiation ", *British Journal of Radiology*, new ser., 25 (1950), 550-552.

GENE RESEARCH OF THE 20ᵗʰ CENTURY REFLECTED BY THE NOBEL PRIZES FOR CHEMISTRY

Claudia LENZNER - Alfred NEUBAUER

One can consider the rise of molecular biology and genetics from different points of views. Recent works have often concentrated on the role of the information theory for the birth of molecular biology, the question of the reduction of life to information[1].

We want to focus on examples of practical work, on experiments which were done in the course of the emergence of these branches of science. One glance at the list of Nobel Prize winners in Chemistry shows that the awarded chemical works brought major contributions in gene research.

Max Delbrück (1978) :

We now use chemistry to shuffle genes,
Use plasmids to move man's to beans,
Or rats to microbes, flies to fleas,
Or yeast's to coli, bee's to peas,
All this is based on Watson's-Crick's
Phantastic double helix, plus some tricks
That others added to this play
and add still more from day to day.

Even if one has to mention that Max Delbrück, who counts as one of the founders of molecular biology, underestimated for many years the contribution chemistry could have and obviously had in genetic research. But it is also clear that classical genetics dealing with the phenotypic description of inherited characters, with laws of transmission and with population studies, was not a question of chemical compounds[2].

---

1. L.E. Kay, " Wer schrieb das Buch des Lebens ? Information und Transformation der Molekularbiologie ", in *Objekte, Differenzen, Konjunkturen : Experimentalsysteme im historischen Kontext,* Hrsg. M. Hagner, Berlin, Akademischer Verlag, 1994.

2. P. Fischer, *Licht und Leben, Ein Bericht über Max-Delbrück, den Wegbereiter der Molekularbiologie,* Konstanz, Universitätsverlag, 1985.

The classical geneticist Thomas Hunt Morgan described, in the mid-thirties, a gene as a point on the line of a chromosome and — as it was mentioned by Judson — " pushed aside any reflections on the physical nature of the genes as premature or unnecessary "[3].

The term " gene " changed a lot in the course of the 20th century. One modern definition for " gene " is the following : A gene is a defined nucleotide sequence in the DNA coding for distinct inheritable determined structures or functions of an organism[4].

The physical nature of genes was determined by the moleculargenetical experiments by Avery and co-workers published in 1944 and by Hershey and Chase 1952, stating that the genetic material refers to DNA and not, as was widely believed before — to proteins[5].

Concerning the final elucidation of the chemical constitution of nucleic acids, the English organic chemist Alexander Todd, Nobel Prize winner of 1957, was instrumental in that field at the end of the forties[6]. His path led him in the thirties — by investigating nucleotide-coenzymes — into the field of nucleotide chemistry. In his autobiography Todd wrote : " Already in 1938 when I started work in this field I was of the opinion that nucleic acids might be involved in the transmission of hereditary characteristics as had been suggested by the earlier work of Griffith on pneumococcal transformation "[7]. Todd, who meanwhile held the professorship for organic chemistry at the University of Cambridge, England, was therefore fascinated by Avery's discovery that the genetic material consists of DNA. So even if, for different reasons, Avery's work did not convince everybody and gave rise to strong discussions in Cambridge about the nature of the carriers of hereditary characteristics, Todd decided to examine the constitution of DNA.

When Todd started to deal with nucleic acids, the so-called tetra nucleotide-theory postulated by Levene represented the dominating idea about the structure of nucleic acids. This theory described the basic structure unit of nucleic acids to be tetra nucleotides with each of the four different nucleotides forming colloidal aggregates in solution.

In 1951 the experiments performed by Todd and Chargaff led to the rejection of the tetra nucleotide-theory. It became clear : Nucleic acids are macromolecules and not aggregated tetra nucleotides. The research of Todd and co-

3. H.F. Judson, *The eighth day of creation : Makers of the revolution in biology*, New York, 1979.

4. *Herder-Lexikon der Biochemie und Molekularbiologie*, Heidelberg, Berlin, Oxford, Akademischer Verlag, 1995, 2. Band.

5. O.T. Avery, C.M. MacLeod and M. McCarty, *J. exp. Med.*, 79 (1944), 137-158 ; A.D. Hershey and M. Chase, *J. gen. Physiol.*, 36 (1952), 39-56 .

6. *Les prix Nobel, The Nobel prizes* — year books of the Nobel-foundation.

7. A. Todd, *A time to remember. The autobiography of a chemist*, Cambridge, 1983, 78-97 .

workers led to the definitive evidence that ribo- and desoxyribonucleic acids are linear 3'-5'- polynucleotides[8].

Two years later the phage geneticist Watson and the protein structure analyst Crick, being newcomers in the field of DNA research, were working at the Cavendish Laboratory of the University of Cambridge, England. They proposed the double helix conformation for the DNA. Their idea was based on the results of Todd and on X-ray crystallographic analyses of Rosalind Franklin and Maurice Wilkins from the King's College in London. This structural chemical work that revolutionized biology was awarded with the Nobel Prize for Medicine/ Physiology for Watson, Crick and Wilkins in 1962[9].

After Todd, it took nearly 25 years for another Nobel prize in Chemistry which honoured moleculargenetical important work : In 1980 the British Frederick Sanger and the two US-Americans Walter Gilbert and Paul Berg shared the Prize[10]. Sanger and Gilbert, independently from each other, gained merits in developing methods for the determination of nucleotide sequences in nucleic acids. The biochemist Paul Berg is considered as one of the founders of the recombination technique for DNA.

Frederick Sanger worked all his life on the structures of biological and medically important macromolecules. He had already got the Nobel Prize for Chemistry in 1958 for the first determination of the amino acid sequence of a protein, namely insulin.

Ever since the mid-fifties there existed many contacts between Sanger's group at the Biochemical Institute and the group of Perutz at the Cavendish Laboratory in Cambridge, UK ; and when Perutz started to negotiate for a new Laboratory for his group Sanger was already involved in these plans[11].

In 1962 the two groups started to work in the newly built Laboratory of Molecular biology in Cambridge and came in much closer contact to each other. To quote Sanger : " In this atmosphere I soon became interested in nucleic acids "[12]. Additionally, Crick, being a good friend of Sanger, turned his attention towards DNA.

As the friendship with Crick was to be an important stimulus for Sanger's work, so was Watson's friendship with Gilbert. Watson, who was meanwhile

8. A. Todd and D.M. Brown, " Nucleotides, Part X : Some observations on the structure and chemical behavior of the nucleic acids ", *Journal of the Chemical Society,* (1952).

9. J.D. Watson and F.H.C. Crick, " Genetic implications of the structure of deoxyribonucleic acid ", *Nature,* 171, 1953, 737-738 .

10. *Les prix Nobel, The Nobel prizes* — year books of the Nobel-foundation.

11. S. De Chadarevian, " Molekularbiologie. Disziplin oder transdisziplinäre Bewegung ", *Biol. Zent.,* bl. 113, 1994, 211-217. ; S. De Chadarevian, " Architektur der Proteine. Strukturforschung am Laboratory of Molecular biology in Cambridge ", in *Objekte, Differenzen, Konjunkturen : Experimentalsysteme im historischen Kontext,* Hrsg. M. Hagner, Berlin, Akademischer Verlag, 1994.

12. *Les prix Nobel, The Nobel prizes* — year books of the Nobel-foundation.

working at the Harvard-University in the American Cambridge woke up Walter Gilbert's interest in molecular biology : " Watson was able to intrigue me with the mysteries of molecular biology "[13]. Under the influence of Watson the theoretical physicist Gilbert converted into an experimental molecular biologist.

While Sanger solved the problem of DNA sequencing mainly with the development of an enzymatic method — a controlled interruption of replication, Gilbert created a sequencing system based on the use of chemical reagents[14]. However, the specific break of a sequence after a determined nucleotide — as the basic principle of the two approaches is comparable. Sanger and co-workers analysed the first total sequence of the genome of an organism in 1977. They determined the 5386 nucleotide sequence of the phage YX 174.

The biochemist Paul Berg from Stanford University, USA, succeeded for the first time in combining DNA-sequences from different species at the beginning of the seventies[15]. His research results revolutionized the genetical chemistry, and opened the doors for the creation of a new industry-genetic engineering.

Berg himself quickly realized the potential dangers of the new technique and suggested ethical debates about genetic engineering. The conference of Asilomar in 1975 united for the first time scientists to discuss the risks of recombinant DNA methods and led to binding security guidelines.

Not only structural analytical achievements or the development of methods were awarded by Nobel Prizes but also the elucidation of the structures of complicated biological molecule complexes consisting of proteins and nucleic acids. It was the South-African British scientist Aaron Klug who had the honour to get the Nobel Prize in 1982 for his research results about the three-dimensional structure of the tobacco mosaic virus and the structure of chromatin.

The X-ray analysts Klug started his research of viruses in 1954 when he succeeded to become a co-worker of Bernal at the Birkbeck College in London. Bernal had begun the X-ray analysis of proteins as well as of nucleic acids already in 1936. It was there, that Klug met Rosalind Franklin, who, after working on DNA, now dealt with the elucidation of the structure of the TMV. Klug will be involved in these works, and he and Holmes go on in this field after the early death of Rosalind Franklin in 1958. The research of Klug and

13. *Les prix Nobel, The Nobel prizes* — year books of the Nobel-foundation.

14. F. Sanger, S. Nicklen and A.R. Coulson, " DNA sequencing with chain-terminating inhibitors ", *Proceedings of the National Academy of Sciences* (USA), 74, (1977), 5463-5467 ; W. Gilbert and A.M. Maxam, " A new method for sequencing DNA ", *Proceedings of the National Academy of Sciences* (USA), 74, 1977, 560-564.

15. P. Berg, " A biochemical method for inserting new genetic information into SV40 DNA : Circular SV40 DNA molecules containing lambda phage genes and the galactose operon of E. coli ", *Proceedings of the National Academy of Sciences* (USA), vol. 69 (1972).

co-workers about the TMV led to basic results on the structure of the protein-doubledisc in the virus, for example the interaction of this doubledisc with the ribonucleic acid of the virus as a part of the self-organisation of the TMV[16].

In 1962, a group of crystallographs from the Birkbeck College, amongst them Klug, was integrated into the Laboratory of Molecular biology in Cambridge in the course of its new occupation. Klug was enthusiastic about this " unique environment of intellectual and technical sophistication "[17]. There, in order to develop models for the structures of biological systems, Klug started to develop his technique based on electron microscopy combined with mathematical manipulation to investigate complex structures.

The years lasting research on the structure of viruses was an important precondition to start analyses on another complicate molecule complex, the chromatin, together with Kornberg in the early seventies.

Chromatin, resembling the total mass of genetical material in the nucleus of the cell, consists of DNA connected with small proteins, the histones. With these proteins the DNA creates repeating units named nucleosomes, which are arranged like pearls on a chain[18].

With these analyses Klug and co-workers contributed to a better understanding of the packing mechanisms of DNA in the cell nucleus and provided the structural and chemical frame for the observation of such events as the replication and the mitosis. Klug said about these findings in his Nobel Lecture that the experiments surprised many biologists because they showed that nearly the whole DNA in the nucleus of the cell is arranged in a very regular structure.

In the previous explanations we mentioned Rosalind Franklin twice in connection with the elucidation of the DNA structure as well as with the TMV. In both cases Franklin gave a major push to the decoding of these fundamental biological structures. It is probably mainly due to her early death that this excellent scientist was not awarded with a Nobel Prize[19].

Sydney Altman and Thomas Cech got the Nobel Prize for Chemistry in 1989 for their independent discoveries of catalytical features of ribonucleic acids, RNA.

The physicist and molecular biologist Altman discovered in 1978 a biocatalyt consisting of a protein and RNA- the RNase P, and that in this molecule complex it is obviously the RNA which acts as catalyt[20]. This research begun

---

16. D.L.D. Caspar, A. Klug, " Physical principles in the construction of regular viruses ", *Cold Spring Harbor Symp. Quant. Biol.,* 27, 1962, 1-24 .

17. *Les prix Nobel, The Nobel prizes* - year books of the Nobel-foundation.

18. R.D. Cornberg, A. Klug, " The nucleosome ", *Sci. Am.* 244 (2), (1981), 52-64.

19. A. Sayre, *Rosalind Franklin and DNA*, New York ; London, W. & W. Norton Company, 1978.

20. S. Altman, M. Baer, C. Guerrier-Takada, A. Viogue, " Enzymatic cleavage of RNA by RNA ", *Trends Biochem. Sci.,* 11, 1986, 515-518 .

again in the English Cambridge when Altman came to the group of Francis Crick and Sydney Brenner in 1969 to work there for two years. The time in this lab gave the basis for the discovery of the RNase P and Altman remembers it as " scientific heaven "[21].

The biophysicist Cech was able to show that ribosomal RNA has the ability to cut it's own nucleotide bonds as well as to reunite them — a process described as self-splicing[22].

With these discoveries, the two scientists came into conflict with the dominant opinion that only proteins can act as enzymes. It was for Cech, Altman and their co-workers a troublesome way full of doubts to become " heretics " against the enzymologists. However, the resistance of the enzymologists against the extension of the term " enzyme " was broken by further experimental evidences.

The discovery of the catalytical abilities of RNA offered new insights into the beginning of life on earth. The answer to the question about the first biological macromolecules on earth is today : Probably RNA molecules were the first biological macromolecules because they combine in them the two fundamental features of life : they can be a genetic code — as the DNA does, and they can act as enzymes — as proteins do.

With the Nobel Prize 1993 the attention of the public was again turned to the field of DNA research. The biochemist Kary Mullis from the USA and the chemist Michael Smith from Canada got the prize together for the development of two important molecular biological methods — the polymerase chain reaction, PCR, and the site-directed mutagenesis.

More by chance than by conscious choice, Smith was so lucky to come as a postdoc in 1956 to the group of the organic chemist Gobind Khorana, a former researcher with Todd. This group dealt with the chemical synthesis of desoxyribonucleotides. Khorana's versatile use of chemical as well as enzymatical methods to solve biological problems impressed the young Smith and influenced his further development. In 1975 he got the opportunity to work for one year in the team of Frederick Sanger. In this time, surely not by chance, Smith's idea of the site directed mutagenesis was born. This method allowed the introduction of a wrong nucleotide into DNA and to cause a specific, site-directed mutation on the DNA level leading to the incorporation of an incorrect amino acid at a specific position in the analyzed protein. This method is technically easily to handle and is today mainly used to clarify the relation between structure and function in proteins. It came into biochemistry as an analytical

---

21. *Les prix Nobel, The Nobel prizes* - year books of the Nobel-foundation.
22. T. Cech, " RNA as an enzyme ", *Sci. Am.* 255, 5 (1986), 64-75.

tool to characterize proteins and is even today undergoing further development up to protein engineering and design[23].

Mullis' path led him from his interest on the synthesis of oligonucleotides to the chemistry of DNA. This interest was awakened after he had finished his thesis when visiting a seminar on the synthesis and cloning of a gene for a hormone at the University of California in San Francisco : " That impressed me. For the first time I realized that significant pieces of DNA could be synthesized chemically and that they were likely to be very exciting. I started studying DNA synthesis in the library. And I started looking for a job making DNA molecules "[24].

Mullis found such a job in the biotechnology company Cetus In California in 1979. The work in the Cetus Company led him to the solution of two annoying problems in the chemistry of DNA : the problems of abundance and distinction. With the polymerase chain reaction, the method discovered by himself in 1983, he could, for instance, multiply a specific sequence region of the three thousand millions nucleotides of the human genome, so that this sequence would be amplified in a sufficient, even visible quantity and that he could determine the precise length of this region even already before[25].

Today, eleven years after the official introduction of PCR in 1986, it has become a standard method in molecular biological laboratories, and the young generation of molecular biologists and geneticists has already grown up with PCR and cannot imagine a research without PCR in the daily work anymore. PCR today is connected for instance with the diagnosis of inheritable or infectious diseases, with solving crimes and mysteries of molecular archaeology.

The reflections made in the frame of this paper may have given the impression that only a small number of personalities essentially advance the scientific development. But this impression has to be modified for most of the investigated achievements. In his book *Cantors Dilemma* the biochemist Carl Djerassi imagines a fictitious Nobel Prize winner in biochemistry on the occasion of the award ceremony who would say : " ...the contribution you have honoured with the Nobel Prize is not the achievement of one or two single persons. It is the culminating point of a long-standing, tiring and apparently unsuccessful research work of many, accompanied from moments of ecstasy... "[26].

23. *Les prix Nobel, The Nobel prizes* - year books of the Nobel-foundation ; M. Smith, " Mutagenesis at specific sites : a summary and persperctives ", *Basic Life Sci.* 20, 1982, 157-160.

24. *Les prix Nobel, The Nobel prizes* - year books of the Nobel-foundation.

25. K. Mullis, F. Faloona, S. Scharf, R. Saiki, G. Horn, H. Erlich, " Specific enzymatic amplification of DNA in vitro : the polymerase chain reaction ", *Cold Spring Harb. Symp. Quant. Biol.*, 51 Pt1, (1986), 263-273 .

26. C. Djerassi, *Cantors Dilemma*, New York, 1989.

But, in focusing on the awarded achievements of the nine Nobel Prize winners it becomes clear that these prizes reflect a series in the basic progress in
genetical research of the 20th century. Important structural-chemical achievements about nucleic acid-protein molecule complexes which required the use
of modern physical methods, were awarded as well as typical chemical or biochemical works on the field of nucleic acids.

As the first category in table 1 shows, there were organic chemists, biochemists and physicists involved in these successes. Not all of them were —
from the point of view of their education — destined for the tasks which had
to be solved. But these scientists were ready to undergo major professional
changes. As one could see, some of these changes were due to personal relationships and influences. Finally, all of them were fascinated by working on
biological problems.

Five of the nine scientists worked for a shorter or longer time in the University of Cambridge, England. Cambridge seems to be a key place, and the scientific institutions of Cambridge with their concentration of excellent scientist
personalities were obviously very important for the development of ideas and
the performing of the awarded achievements. Latour creates for example in
1987 the expression " obligatory passage point " for the Laboratory of Molecular Biology in Cambridge[27].

The concrete cases show that the chemical work which was awarded by
Nobel Prizes introduced new experimental systems in the sense of Hans-Jörg
Rheinberger[28] and tools for the daily work in the laboratories as well as giving
impulses for changes in conceptions in molecular biology. The awarded
achievements constitute a sample which can contribute to illustrate the inter-
twining of chemistry into what became molecular biology.

---

27. B. Latour, *Science in action. How to follow scientists and engineers through society*, Cambridge, Mass., Harvard University Press, 1987.
28. M. Hagner, H.J. Rheinberger, B. Wahrig-Schmidt, *Objekte, Differenzen, Konjunkturen :
Experimentalsysteme im historischen Kontext*, Hrsg. M. Hagner, Berlin, Akademischer Verlag,
1994.

Some data about the Nobel Prize Winners

Nobel Prize Winners	First Training	Nationality	Data of Life	Achievement	Year of Nobel Prize
Todd, A.	Organic Chemi-stry	Great Britain	1907-1997	Nucleotides/Nucleotide-Coenzymes	1957
Sanger, F.	Biochemistry	Great Britain	1918-	Base sequences of nucleic acids	1980
Gilbert, W.	Theoretical Physics	USA	1932-	Base sequences of nucleic acids	1980
Berg, P.	Biochemistry	USA	1926-	Recombinant DNA	1980
Klug, A.	X-ray Structure analysis	South Africa/Great Britain	1926-	Nucleic acid-protein complexes	1982
Cech, T.	Physicochemistry. Biophysics	USA	1947-	Self splicing RNA	1989
Altman, S.	Physics, Biophysics	Kanada/USA	1939-	Enzymatic activity of RNA	1989
Mullis, K.	Biochemistry	USA	1944-	Polymerase chain reaction	1993
Smith, M.	Organic Chemi-stry	Great Britain/Kanada	1932-	Site directed muta-genesis	1993

MOLECULAR UTOPIAS
FROM HORMONES TO GENES AS THE IDEAL DRUG

Christiane SINDING

Medical issues are components of most utopias : controlling hygiene, sexuality, and the general maintaining of good health is one of their major goals. Conversely, a number of physicians and biologists have outlined the social and political scenarios which, they claimed, were rational outcomes of biological theory. These scenarios can be thought as utopian, in the sense that they associate political concerns with a critical appraisal of the present day society and the hope for the construction of a new ideal society. Karl Mannheim described modern utopias as rational or liberal systems whose main feature is faith in the power of human intelligence, with no consideration for reality, particularly the social context of poverty, violence and money[1]. Although utopias are generally considered to be unrealistic dreams, they have a positive function, which is also critical and constructive. On the constructive side, utopian narratives describe a general improvement in the human condition through the elimination of disease and suffering, a longer life, and a promise of pleasure. On the negative side, their involvement in rational projects for the regulation of populations can involve such unwelcome features as eugenics and Malthusianism.

I believe that, at the end of the 19th century, medicine began to construct new powerful therapeutic tools as dual agents. These cured or alleviated diseases, but could also be used as normative tools. The availability of new active molecules allowed biomedical utopias to be partially achieved. Michel Foucault suggested that the reshaping and disciplining of healthy bodies, as well as the regulation of populations, became a major function of medicine at the end of the 18th century[2]. Beside a critical examination of these views, I would

---

1. See P. Ricoeur, *Lectures on Ideology and Utopia*, New York, Columbia University Press, 1986.
2. See " The politics of health in the eighteenth century " in M. Foucault, *Power/Knowledge. Selected Interviews and other Writings, 1972-1977*, Brighton, Harverster, 1980, 177.

argue that such medical " utopias " are not just ideological by-products of scientific theory, but are in fact major components of them.

THE CONSTRUCTION OF HORMONES AS DUAL AGENTS :
FROM DRUGS TO NORMATIVE TOOLS

Hormones were first constructed in the study of *disease*[3] ; documents were produced about the clinical and anatomical aspects of peculiar chronic and often lethal diseases ; representations of the anatomical structures were constructed, and called " endocrine glands " (1909) . Some patients were given chemical extracts of the anatomical structures which appeared to be involved in the disease and were saved. These therapeutical extracts were called " internal secretions " and said to contain specific molecules called " hormones " (1905). A general representation of endocrine diseases was outlined : they occurred when endocrine glands and their internal secretions were destroyed, so that " juices " produced by the glands involved had to be extracted from animals, and given to patients. Excess of hormone secretion could occur in some cases, usually produced by a hypertrophic or tumoral gland.

Thus, endocrinology was constructed on practical needs, mainly medical. Hormones took on, at least two meanings, after 1950, according to the networks involved in hormone research. In medical networks hormones were still studied as replacement drugs for hormonal deficiencies. Biochemists began to synthesize hormones as early as 1927. But physiologists represented the endocrine system as a highly organized, hierarchical system. The brain and the hypothalamo-hypophyseal area were thought of as the " leader " of the system, sending signals and orders to " peripheral glands " (*target* organs). *Communication* and *regulation* became the main metaphors used in endocrinology.

This representational device had been constructed at the very beginning of the century by Ernest Starling (1866-1927), a physician who devoted his career to physiological research. At this time very little was known about internal secretions, but thyroid extract had been shown in 1919 to alleviate hypothyroidism and adrenal extracts to have a marked hypertensive activity in animals. Starling himself, with Bayliss, had shown that the duodenum secreted a chemical substance which elicited the excretion of pancreatic digestive juices. Putting together those observations and others made on the biological effects of other well-known drugs, Starling ascribed to internal secretions and drugs the

3. For more information see M. Borell, " Origins of the Hormone Concept : Internal Secretions and Physiological Research 1889-1905 ", *Ph. D. Diss.*, Yale University, 1972 ; M. Borell, " Organotherapy, British physiology and discovery of internal secretions ", *Journal of the History of Biology*, 9 (1976), 236-268 ; C. Sinding, *Le Clinicien et le Chercheur. Des Grandes Maladies de Carence à la Médecine Moléculaire, 1880-1980*, Paris, Presses Universitaires de France, 1991, 284 p.

function of " chemical messengers ". He stated that natural chemical messengers served to *integrate* and *control* the various functions of the body and added : " If a mutual control of the different functions of the body be lately determined by the production of definite chemical substances in the body, the discovery of the nature of these substances would enable us *to interpose at any desired phase* in these functions, and so to acquire an *absolute control* over the workings of the human body "[4].

Seventeen years later Starling gave his lecture on the " Wisdom of the Body ", in which he compared the living body to a machine. This was not new and many had done so before and after him. But to him the body was an intelligent machine endowed with wisdom, as implied by the title of his lecture[5]. He thus introduced teleology into his explicitly mechanistic view of life. He also insisted on the miracles of science, whereas at the same time he explained the need for long and patient investigations of physiological functions. He emphasized the fact that " Pure " science was autonomous and disinterested, but also stated that science was in charge of improving the human condition. Such contradiction and ambivalence are at the heart of all utopias, which combine dreams with rationality, and the escape from reality with the construction of a new world. Similarly, Starling shifted easily from the curing function of medicine to its normative and regulating one, just as he played with the double status of " drugs/chemical messengers ", which was analyzed above. He recalled the sentence from his previous lecture (1905) quoted above, and commented : " I think (…) we have made considerable progress towards the realization of this power of control which is the ultimate goal of medical science "[6].

Going further he stated : " It would, indeed, be an advantage if we could postpone the slowly increasing incapacity which affects us all after a certain age has been passed. Pleasant as it would be to ourselves, it would still be more valuable to an old community such as ours, where the arrival of men in places of rule and responsibility coincides, as a rule, with the epoch at which their powers are beginning to diminish "[7].

The myth of rejuvenation and eternal youth that British physicians had criticized so much when Brown-Séquard published his famous papers, came back

4. E.H. Starling, " The Chemical Correlation of the Functions of the Body ", *The Croonian Lectures, Lancet*, (1905), pt. 2, 339-341.

5. E.H. Starling, " The Harveian Oration on the Wisdom of the Body ", *Lancet*, (1923), pt. 2, 865-869.

6. E.H. Starling, *op. cit.* (cf. fn. 5), 869.

7. E.H. Starling, *op. cit.* (cf. fn. 5), 869. Starling had been deeply shocked by the loss of a generation of young men, and thought that the issue of the war could easily have been a disaster. He had stated that " The allies had neglected science and misconceived or despised education " and criticized the " astounding and disastrous ignorance of the most elementary scientific facts displayed by the members of the government ". See C.B. Chapman, " Ernest Starling. The Clinician Physiologist ", *Annals of Internal Medicine Sup.*, 2 (1962), 1-43.

in Starling's discourse[8]. I have shown elsewhere that this myth has always per-meated endocrinology[9]. The idea that a special class of people endowed with more wisdom than the rest, should rule society is also a common component of utopias, as well as the critical appraisal of the present society.

TREATING HEALTHY PEOPLE : UTOPIAS REALIZED

As advocated by Starling, as soon as hormones were isolated and synthe-sized — most of them between 1920 and 1940 — they were not only used for curing diseases but also for interfering with physiological processes in *healthy people*. Sex hormones were among the first to be used this way, and the med-icalization of menopause and menstruation had opened up an enormous market for almost all women as early as the 1930s[10]. But the most interesting mole-cule constructed as a regulating tool for populations is the contraceptive " pill ". Borell has shown that contraception in the 1920s and the 1930s was linked above all to the idea that birth control would eliminate the social prob-lems and evils thought to come from excessive fertility[11]. Adele Clarke has analyzed the complex networks involved in the construction of the pill and has shown that, among the various actors involved in the question of birth control, scientists redirected contraceptive research from practical to scientific meth-ods. Among the latter, the pill was the most attractive, because it could easily be constructed as a simple, effective and universal method[12]. The contraceptive " pill " was deliberately designed to control fertility but other substances first designed to palliate endocrine disorders, have been and are used in healthy people. Growth hormone is given to healthy children with short stature ; vari-ous synthetic hormone antagonists are prescribed for children who are too tall, in order to stop their growth ; male hormones are prescribed for athletes ; it has been suggested that antagonists of male hormones be given to men guilty of sexual abuse (" chemical castration ") ; thyroid hormone is often prescribed for obese people ; insulin has been used to cause comas in patients suffering from mental disorders and drug addicts. Brain chemical messengers have been dis-covered and synthesized, whereas older drugs such as tranquillizers have been shown to counteract the physiological effects of those messengers. Biochem-ists are able to synthesize molecules which differ slightly from the " natural "

8. See C. Sinding, *Une utopie médicale. La sagesse du corps d'Ernest Starling*, Arles, INSERM/Actes-Sud, 1989, 99 p.

9. The recent so-called pill of youth, as constructed by E.E. Baulieu, confirms this view.

10. Nelly Oudshoorn, " United we Stand : the Pharmaceutical Industry, Laboratory, and Clinic in the Development of Sex Hormones into Scientific Drugs, 1920-1940 ", *Science, Technology, and Human Values*, 18 (1993), 5-24.

11. Merriley Borell , " Organotherapy and the Emergence of Reproductive Endocrinology ", *Journal of the History of Biology*, 8 (1985), 1-30.

12. Adele Clarke, *Disciplining Reproduction : American Life Scientists and 'the Problem of Sex'*, Berkeley, University of California Press, 1998, 410 p.

molecules and have different effects, ranging from very similar to opposite. Solomon H. Snyder, who played an important role in the discovery of endorphins (brain " morphins ") wrote : " Indeed, the ability to use modern molecular strategies to fashion extremely potent and selective agents suggests that a whole generation of more useful drugs about to be born. New drugs might play a useful role in influencing emotions and behaviors in ways we cannot imagine today "[13].

Genes have now replaced hormones or other molecules as the best tools for controlling diseases in contemporary medical discourse. Although gene therapy has not been a great success so far, scientists often speak as if it was really working. Because, for technical reasons, " bad " genes cannot easily be replaced by " good " ones, the idea of gene therapy has been reshaped : the gene is now considered to be the best medication for non genetic diseases, which can or should be cured or palliated by protein drugs. In the majority of cases, genes that direct the synthesis of these drugs are injected into patient by using viral vectors. This new form of gene therapy is still complicated and expansive, and is far from being a success. Despite these shortcomings, this new kind of therapy is regarded by many geneticists as the best medication for the future. Thus the latest " magic bullet " is the gene, with the same dual potentiality as hormones and chemical messengers to cure disease and to transform human beings.

CONCLUSION

As mentioned above, Foucault has argued that medicine was no longer confined to curing or alleviating diseases after the Enlightenment, but was in charge of the knowledge of the healthy man and a definition of a model man[14]. He pointed out that the doctor became " the great advisor and expert (...) in correcting and improving the social " body " and maintaining it in a permanent state of health "[15]. However Foucault had focused his studies mainly on techniques for disciplining bodies, apparatuses of health or for governing sexuality and reproduction, without paying much attention to the new powers of biology and biomedicine. Very early in the century, some physicians took advantage of the potent effects of some glandular extracts or vitamins to plead for a transformation of men and society based on the use of these new drugs. Those goals

    13. Solomon H. Snyder, *Brainstorming. The Science and Politics of Opiate Research*, Cambridge, Harvard University Press, 1989, 208 p., 187.
    14. M. Foucault, *Birth of the Clinic : An Archeology of Medical Perception,* London, Tavistock, 1973. He wrote for instance :" In the ordering of human existence (medicine) assumes a normative posture, which authorizes it not simply to distribute advice as to healthy life, but also to dictate the standards for physical and moral relations of the individual and of the society in which he lives ". p. 34.
    15. See " The politics of health in the eighteenth century ", in M. Foucault, *Power/Knowledge. Selected Interviews and other Writings, 1972-1977*, Brighton, Harverster, 1980, 177.

were different from those of eugenics, in the sense that eugenics seeks to improve the human race, whereas those physicians were trying to act on individuals. However the " pill " can be used in both ways, allowing women to gain more " freedom ", or to allow the makers of health policy to improve their control over the population. Lily Kay has stated that " The eugenics goals, which had informed the design of the molecular biology program and had been attenuated by the lessons of the Holocaust, revived by the late 1950s "[16]. I have tried to show that the " molecular vision of life " is not restricted to molecular biology, but that it has permeated all biomedical sciences since the end of the 19th century. Therapeutic molecules have often been constructed as flexible tools which allow the treatment of disease and the control and regulation of individual bodies or an entire population.

16. Lily E. Kay, *The Molecular Vision of Life. Caltech, the Rockefeller Foundation, and the Rise of New Biology*, New York, Oxford University Press, 1993, 304 p., 277.

THEORETICAL BIOLOGY IN PHYSICO-CHEMICAL AND BIOLOGICAL CONTEXTS[1]

Vera GUTINA

First of all, what do we mean by the term " theoretical biology " ? The long-term discussion on this topic resulted in a rather consistent understanding of theoretical biology as a field that is distinct from general biology and that integrates the most important data from various fields of biology. In comparison with general biology, the task of theoretical biology is understood as the creation of a unified theory of living nature based on the knowledge about all levels of the organisation of living, and on the knowledge about the laws of the origin of life, of its functioning, development, and evolution. The theoretical physics is widely regarded as a model here in terms of exactness, fundamental character and axiomatics of its ideas about one subject and phenomenon.

In fact, we have the following situation at present. We know that, on the one hand, some authors completely deny even one principal possibility of creating theoretical biology because of the multilevel character of such a self-organizing, artificially not reproducible, developing system penetrated by information flows as is living nature. For example, the great Russian biologist who began his scientific activity with the physicist M. Delbrück, wrote in 1983 : " Theoretical biology is absent today, or was absent until the very last time, because there are not (or there were not until the very last time) the general natural-historical biological principles, which could be comparable with those, that already since a long-term, beginning by the 18th century, exist in physics now "[2].

On the other hand, however, we have two positions in the interpretation of this problem. According to the first, theoretical biology can be constructed on the basis of the so-called " physicalist " programmes. In other words, it should

---

1. English version revised by B. Hoppe, München, BRD.
2. N.V. Timofeeff-Ressovskiy, " Genetics, Evolution and Theoretical Biology ", in *Idem, The Biospheric Meditation*, Moscow, 1966, 77-82 (in Russ.).

synthesize the main lines of the mathematical (cybernetic) and physico-chemical approaches, including thermodynamics, theory of systems, information's theory and other and already known physical laws and principles or the new ones.

According to the second position biology should, in a sense, rely upon itself on realizing a synthesis of knowledge about such a unique phenomenon, as life. It should be mentioned in connection with it, that developing for more than a millennium, the theoretical thought in biology followed distinctively different patterns in comparison with the development of theoretical physics or chemistry.

In the history of the first, i.e. the physico-chemical trend, the main ideas may be differentiated into two paradigms — the chemical and the physical one. Let's start with the first one. By characterizing the chemical paradigm we must note that chemistry, but not physics had a historical priority in the development of the physico-chemical trend. As it has been shown (Table 1), the F. Engels' legendary aphorism — " the life is a mode of existence of protein bodies " — had become a dogma, in any case, in soviet science, predetermining for many years the leadership of chemistry in the interpretation of the essential characteristics of life. But also two purely chemical factors forced chemistry's leadership in the interpretation of the life's essence. These were :

1. The chemical analysis and synthesis of the " building blocks " of living matter and

2. the deciphering of the biochemical reactions of metabolism. It made it possible for A.J. Lehninger to say : " Biology is a kind of super-chemistry " and " A cell is a machine... The living... organisms obey the systems that we may call molecular logic of living "[3].

All the four factors, including the immense successes of molecular-biological reductionism, have contributed to the strengthening of the chemical paradigm. But it's known that during experimental, exactly analytical molecular-biological and biochemical researches " life is dying away ". Its objects are becoming the separate fragments of the living matter of different organisation's levels. The same situation takes place by the theoretical model of the essence of life on its low, i.e. fundamental basic level. It's a source of the greatest simplification and schematization of living entirety.

In the context of the physical paradigm might be differentiated three notion's groups (Table 1). The first group is represented by ideas, in accordance with all physico-chemical laws and principles, including the Second Law of Thermodynamics, are completely adapted to explanation of the life, its functioning and evolution. With respect to the second group of my classification the scientists, belonging to this group, tend to create new physico-chemi-

3. A.J. Lehninger, *Biochemistry. The Molecular Basis of Cell Structure and Functions*, Moscow, 1974, 10 (in Russ.).

cal principles and even laws by introducing deep modifications into the already existing regularities with a purpose to explain and to model the essence more exactly and correctly, but the most important, the origin of life. A central question of this trend is a legitimate application of the main ideas of the Second Thermodynamics' Law " energy " and " entropy " to the life. The main reason for the elaboration of this problem is the fact, that the world of living organized systems, i.e. organisms, thermodynamically (concepts " energy " and " entropy ") behave in " a different way ", because the evolution of living leads to the involvement of more complex systems. but not to " the degradation, equilibrium and not to the stationary conditions ", in accordance with ideas of E. Bauer[4] — one of the representatives of this trend, i.e. of the second group. Following his postulates, Bauer became able to overcome this significant contradiction by formulating a certain " principle ". According to him, even on the molecular level of living there exists some non-traditional " thermodynamic stable non-equilibrium, achieved on the cost of the free energy of molecules, working against the equilibrium " (Bauer).

After the elaboration of the general theory of systems by L. von Bertalanffy[5] today three types of systems are distinguished :

1. the opened systems, i.e. changing with environment both with energy and matter ;

2. the reserved systems, i.e. changing with environment by energy only ;

3. the closed, or isolated systems, changing with environment neither by energy, nor by matter.

TABLE 1

PHYSICO-CHEMICAL TRENDS IN THEORETICAL BIOLOGY

The chemical paradigm	The physical paradigm
1. " The life is a mode of existence of protein bodies " (F. Engels' thesis).	1. Physico-chemical laws and principles are completely adapted to the explanation of life and its evolution (L.A. Blumenfeld, M.V. Volkenstein).
2. The analysis and synthesis of the " building blocks " of living matter.	2. Already known physico-chemical laws should be modified and adapted to living nature or elaborated as new principles and laws (E. Bauer, E. Schrödinger, L. Brillouin, A. Dükrock, I. Prigogine et al., G. Haken).

4. E.C. Bauer, *Theoretical Biology*, Moscow, 1935, 8 (in Russ.).
5. L. von Bertalanffy, *Theoretische Biologie,* H. 1, Berlin, 1931.

3. The deciphering of the composition, structure and metabolism of living organisms. " The biology is the superchemistry ". " The cell - is a machine " (A. Lehninger).

3. There are two independent physico-chemical laws : one was adapted to the explanation and deciphering of living nature, the other — to the non-living one (K.C. Trinsher, G. Schaefer *et al.*).

4. Molecular-biological reductionism : the deduction of the entire living system (organism) to the sum of its parts.

This scientific achievement served as conceptual basis for another representative of this second group — E. Schrödinger[6] for, at first, subsequent modelling of life on its fundamental level and, at second, for the following development of theoretical biology. According to Schrödinger's views, origin and organisation of living are consequences of the entropy with the inverse sign. Therefore, it is itself the measure of molecular order. However, as Schrödinger has admitted, " this inversion of the sign of entropy doesn't mean that the Second Law of Thermodynamics is adequate for the description of the living because of the mechanism, on which the entire statistical physics is based, doesn't work for this case "[7]. This means, according to Schrödinger, that we may assume that the living matter obeys a new type of physical regularity of non-statistical nature.

A consequence of this thesis was a postulation of the mechanism of life's existence : an unbroken extraction of the regularity from the environment. It is achieved by constant functioning of the dichotomy " system / environment " because of the organism is obeyed by the property " to drink a regularity from suitable environment ", according to Schrödinger's expression. As a result of this concept of Schrödinger a new base was constructed for the generation of new principles for the interpretation of the " essential characteristics " of living in the framework of physical paradigm.

In the same camp with E. Schrödinger was L. Brillouin[8], another remarkable figure who attempted to link the classical thermodynamics to the theory of information, interpreting the Second Law of Thermodynamics in the context of irreversibility of evolution, i.e. in connection with the factor of time. Besides

6. E. Schrödinger, *What is Life ? The Physical Aspect of the Living Cell,* Moscow, 1977 (in Russ.).

7. *Idem.*

8. L. Brillouin, " The Thermodynamics - Cybernetics Life ", in *Cybernetics. A current state,* Moscow, 1980 (in Russ.).

that he attempted to resolve the problem of the relationship between this Law and the facts of variability of entropy. As consequence has been formulated a new notion — " negentropy ", i.e. entropy with negative sign, which became a most adaptable explanation of the relationship between living nature and the Second Law of Thermodynamics. It is necessary to remark that just this Brillouin's idea has been used by Schrödinger for the unique integration of two notions — entropy namely, and regularity of living matter.

For the same reason the finding of new physical principles and laws which would be capable to explain the living on its fundamental basic level A. Dükrock[9] has estimated as a base for the creating of new physics suitable for describing not only statistical, but also organized, asymmetrical and opened systems with the reverse connections. That's why Dükrock has appealed to the cybernetics that, at his opinion, despite of its capacity for modelling the isolated, closed systems, it might, according to Dükrock, describe a structure and functioning of " the factor regularity of living nature ". It may seem that because of these conditions everything predetermining, the emergence of the theory of structural self-organization in order to model the life and pre-life simply couldn't appear. And this mission has been realized by a Brussels' school of naturalists, represented by T. de Donder (1931), E. Yantsch[10], I. Prigogine[11] and his colleagues — P. Glansdorff, N. Nicolis and others and in the same time (1970-1980) by G. Haken — one of the creators of the synergetic's theory[12].

By the way, it's necessary to remark that the term " self-organization " has been proposed by the physicist W. Asby in 1947. But this concept was deeply rooted just in the biology, and in theological and gnoseological respects it became indissoluble connected with a biological concept of " organization ". But it is absolutely true, according to H. Haas[13], and it is very significant for the development of theoretical biology, that one must just distinguish two different concepts : both the concept of " self-organization " and " self-structuralization ". Why ?

The fact is that a concept of " organization " terminologically and ontologically emerged from the term " organ ". Just because a concept of " self-organization " should be relevant only " to describe living systems " (Haas), i.e. living nature. At the same time the concept of " self-structuralization " has

9. A. Dükrock, " The physics and cybernetics ", *The Cybernelics. The Results of its Development,* Moscow, 1980 (in Russ.).

10. E. Yantsch, " The Self-Organizing Universe ", *Scientific and human Implications of the Emerging Paradigm of Evolution,* Oxford 1980, 343 p.

11. I. Prigogine, " The Statistical Interpretation of Non-equilibrium Entropy ", in *The Boltzmann Equation. Theory and application,* New York, Berlin, 1973, 401-450.

12. G. Haken, *Synergetics. An Introduction,* Berlin, 1977, 350 p.

13. H. Haas, *Natur und Begriff. Fachdidaktische Studien des Assoziationsraums biologischer Begriffe mit besonderem Schwerpunkt auf Chaos und Ordnung,* Hamburg, Diss. Univ., FB Erziehungswiss, 1995.

a sense solely with respect to the " inanimate nature ", i.e. non-living systems. According to G. Schaefer[14], in this connection a question about the fundamental resemblance and distinction of both these phenomena, i.e. " living systems " and " non-living systems " raised to the level of conceptual understanding, including four concepts — Life, Order, Energy and Information. A correlation between them is today one of the main problems of theoretical biology, developing in the framework of a physical paradigm (a third group).

But let's return to the Brussels' school. In the centre of its meditation was the integration of the Thermodynamic Carnot-Clausius' principle that postulates the process of evolution in accordance with the Second Law of Thermodynamics and understands it as a degradation and destruction of an initial structure with the principle of biological evolution associated with the development and sophistication of biological systems. In other words, in the framework of the already existing physical laws a theory was created, which is comparable with the theory of biological evolution and which describes the biological systems as distant from the thermodynamic equilibrium of open systems, the non-equilibrium of which is sustained by the constant inflow of constantly dissipating energy. In this way, the term " dissipative systems " has gained currency for such systems : the theory of non-equilibrium thermodynamics has been created. This theory also postulated another important characteristic of dissipative structures their non-linear character.

In the context of synergetics the integration between physics and biology, and the contradiction between the Second Law of Thermodynamics and the regularities in the structure of the objects of the organic world should be removed. The universal description of co-operative interaction in physical, chemical, biological, and even socio-cultural phenomena is the goal of this field of knowledge, according to its leader G. Haken. It could seem that the physico-chemical line in theoretical biology gained a new impetus for its development, and the theoretical biology got a new basis for the construction. But the euphoria has faded rather soon, and the scepticism has made its appearance. But why ?

First, because the acute question of the role of selection on the prebiological level, the possibility of which Th. Dobzanskiy regarded as " senseless ", was addressed again. Second, the question about the role of various forms of selection " responsible " for the origination of the unique biodiversity of living nature absolutely cannot be resolved in principle in the framework both of the physical and chemical paradigms. And finally and it's most important, the concept of the universal character of non-equilibrium thermodynamics, including synergetic, and even triumphs of molecular-biological reductionism, which encouraged the deciphering of functioning of living on its fundamental level,

14. G. Schaefer, *The Synergetic Notion about the Concept of the Quadrangle " life-order-energy-information "*, Kiel, 1966.

haven't approached the understanding resting unresolved and today of basic riddle of life — the problem of its origin. F. Redi's aphorism — *omne vivum e vivo* — can sound rather loudly anew...

It is not an accident in this situation that the third group of ideas in the framework of the physical paradigm has manifested itself very strongly, this means, the development of the idea, according to which " all physical laws and principles of living matter have an independent nature ". And this means, in accordance with the Russian scientist K.S. Trinsher[15], one of the most typical representatives of the third group, that " living molecules [i.e. the molecules of animate matter — V.G.] are everything that cannot emerge from the known inanimate matter, because the physics of living matter has an independent nature ". This courageous thesis required a more deep explanation and Trinsher had written : " In the fundament of life there are the processes of the overcoming of the thermal chaos, emerging in the living system itself by the life's temperature. These processes are antientropic and are contradictory to the Second Law of Thermodynamics ". About G. Schaefer's ideas concerning the fundamental discrepancy of two phenomena — " living systems " and " non-living systems " — we have spoken already above.

Thus the physico-chemical background of theoretical biology got nothing for the conceptual interpretation and the model of existence of life on the basic level and for its origin. Therefore, it is small wonder that the biologists and even philosophers began to unite their rows anew, in order to create a scientific background for erecting the building of theoretical biology. But meanwhile an ancient competition between mechanicizm and vitalism is being continued only acquiring new vestments.

One of the many attempts to resolve the problem of creating a scientific background of theoretical biology on the base of a biological paradigm was a concept of structural organization's levels of living matter. One must say that the sources of these attempts are very early : they had proceeded from the idea of a discrete structure of living entirety (the works of M.J. Schleiden, H. Spencer, E. Haeckel and others). However, in the most complete form this concept has been created by two philosophers — H.Ch. Brawn and R.W. Sellars (1920). Later on (in 1940), this concept was transformed into the theory of integrative organization's levels and was clearly interpreted by A. Novikoff (USA) in 1945 and then in 1950 by W. Kremianskiy and K.M. Hayloff (USSR). In 1950-1960 the completed concept of organization's levels of life has been stated by N.V. Timofeef-Ressovskiy[16]. He distinguished four levels : 1. the molecular-genetic, 2. the ontogenetic one, 3. that of population,

15. K.S.M. Trinsher, " Could it be artificially to create the living ", *The Questions of philosophy*, 9 (1965), 120-127 (in Russ.).

16. N.V. Timofeeff-Ressovskiy, " The organization's concept of living ", in *Idem, The Biospheric Meditation*, Moscow, 1966, 186-201 (in Russ.).

and 4. the " biogeokoinosic ", or biospheric one. But all these attempts have not been completed with full success : this was not a true biological paradigm for the building of theoretical biology.

Then (1970-1980) the discussing had been started from the question about which of the three " images " of biology — the classical, or the naturalistic, physico-chemical or the evolutionary one[17] — could serve as the most adequate way and as a platform for the construction of the theoretical biology on the base of biological paradigm. Among these three " images ", the evolutionary biology had been recognized as the most suitable as long as it can serve as the most vast platform to carry out the interdisciplinary synthesis of the data, obtained by the study of the living on all levels of its organisation — from molecular to biospheric — with a purpose of finding the unified regularities in functioning and development of animate nature. But, only about the already known animate nature ! The main problem — how originated this animate nature — remains unresolved. All attempts, scientifically in any case, in frameworks of chemical or physical paradigms to model this mysterious act, finished unsuccessfully at present.

Therefore, we understand why a second paradigm — a biological one — has shaped. Its general idea could be depicted by the words of the great naturalist V.I. Vernadskiy : " There are impenetrable borders between the living natural body and non-living (*kosniy* or *biokosniy*) body ". Since the oldest time such an idea has been qualified as a vitalistic one and all scientists who have shared this idea have been estimated as vitalists. Just in this sphere continued the acute struggle between mechanicizm and vitalism. However, many so-called " vitalists " have shared true naturalistic views. For example could be mentioned a Russian experimenter and thinker I.P. Borodin[18] who had written : " It's indisputable that in living bodies are operating the physical and chemical laws... But current the state of our knowledge about life's phenomena should be premature to support that the known physical laws are rather suitable for the explanation of life ". This had been written in 1893 ! And very remarkable are the following words of Borodin : " The organisms undoubtedly submit to the laws of the eternity of matter and energy's conservation, what, however, didn't exclude the existence of a special, organizing principle " (*ziydilskoe nachalo*)... reasonably using the matter and forces of inanimate nature in order to direct its action to a known purpose — to create and conserve the living organisms ". These words are not absolutely apprehended today as a sign of vitalism : they express only the true state of our knowledge about the condition of life's origin. How these words are consonant to the views of our recent great naturalise and thinker, a founder of molecular biology in Russia V.A. Engel-

---

17. V.N. Gutina, " Three 'Images' of Biology ", in V.I. Kuznetzoff, G.M. Idlis, V.N. Gutina, *Natural science,* Moscow, 1996, 251-277 (in Russ.).

18. I.P. Borodin, " The Protoplasm and Vitalism ", in *The Gods World,* St. Petersburg, 1983 (in Russ.).

hardt[19], who said in 1984 : " As concerns the molecular biology we must declare that as before remained not understood one most radical, greatest task, standing as before : it's essence, nature and origin of life's phenomena ". May be unusually that the creational conception, as we know, gathers the forces again in our days, and it's not without ground ? (Table 2).

TABLE 2

" There are impenetrable borders between the living natural body and non-living " (V.I. Vernadskiy, O. Hertwig, I.P. Borodin, V.A. Engelhardt).

But what is the situation in theoretical biology today ? The fact is that in the framework of a biological paradigm in accordance with the thesis of N.V. Timofeeff-Ressovskiy[20] certain biologists have made attempts to deduce any axioms analogous to physical ones, but characterizing life, and only life, in order to create the foundation for the theoretical biology. One of these attempts has been undertaken by B.M. Mednikoff[21]. He formulated four axiomatic principles, " constituting the life's base ", according to him :

1. The living organisms are formed by the phenotype and genotype, i.e. have the genetic program ;

2. The genetic programs didn't arise anew, but are replicate by the matrix's way ;

3. In the process of replication are inevitable the mistakes on a micro level, the accidental and unpredictable changes of genetic programs ;

4. During the processes of building of the phenotype these changes many times force a possibility of the selection of unitary quantum events on the micro level.

What may assert these axioms ? Yes, they depict the fundamental characteristics of the organic world. But, only on the micro level of its organization ! Might they be applied to the macro levels, i.e. to all organization's levels of living matter, for example, the populations, the organism's communities ? Of course, not.

There is another example to deduce the primary postulates, characterizing the living nature. N.V. Timofeeff-Ressovskiy names two such postulates :

1. natural selection and 2. the so-called " convariant reduplication ", i.e. the hereditary variations on the molecular-genetic level. However, the author himself has acknowledged, that this second postulate is not rather " strict " yet (p. 80)[22].

19. V.A. Engelhardt, " About certain Attributes of Life : the Hierarchy, Integration, 'Recognizing' ", in *The Knowledge of Life's Phenomenon,* Moscow, 1984, 221-248 (in Russ.).
20. *Idem* 1.
21. B.M. Mednikoff, " Axioms of Biology ", *Science and life,* 10 (1982), 35-39 (in Russ.).
22. *Idem* 1.

Besides this, there are a lot of other attempts to construct the base of theoretical biology, i.e. a general theory of living matter[23]. Currently the main trend in this realm is the global interdisciplinary synthesis of theoretical thought in evolutionary (the theories of micro- and macro evolution) and physico-chemical biology, ecology, biogeochemistry etc. We can state that a central place in this searching occupies V.I. Vernadskiy's concept of " living substance ", which, according to Vernadskiy, has existed always and endless. Who created it ? Unknown yet.

Today the naturalistic dogma still dominates. But more and more is growing A. Szent-György's voice, who said in 1978 (Puschino on Oka, Russian) : " The biology is a science of the unbelievability... In the living organisms there are the metabolic reactions which seem like physical ones, but fabulous. But how did origin this life ? It's in general an unresolved puzzle ".

But the modern global environmental ecological, and not only one, but both biopolitical and biotechnological crisis serves a powerful motive for the making in order to realize the need of a science-based conception of the resolution and management over the functioning and evolution of living matter. To resolve this problem is possible only by co-operation of theoretical biology with all natural sciences, with politics, sociology and education, with the entire global knowledge about origin and functioning of the living nature.

23. E.N. Mirzoian, " The Theoretical Biology - Current and Future ", *Izvestia RAS*, Ser. biol., 5 (1993), 774-777 (in Russ.).

PART TWO

THE NEW BIOLOGY OF DEVELOPMENT

FOREWORD

Charles GALPERIN - Scott F. GILBERT

The title of our symposium might appear overly ambitious or presumptuous if we seemed to hint that there were no significant fundamental discoveries in the twentieth century besides those of cell biology and developmental genetics, the two subdisciplines mainly represented here. This present symposium was constructed during the centenary of E.B. Wilson's pathbreaking book *The Cell in Development and Inheritance* (1896), with an eye on his tradition of biological research. We hope that other volumes will follow that will address the equally important areas of biology that were not included in this symposium.

Wilson, of course, attempted to integrate cytology, embryology, and the chromosomal theory of inheritance into a common cellular framework. Theodor Boveri and Claude Bernard were his exemplars. His hopes for synthesis, however, did not come to fruition in his lifetime. It was only in the late 1970s that the synthesis of cytology, embryology and chromosomal theory (now called cell biology, development biology, and genetics, respectively) came into existence. Work which was carried out in Zürich under E. Hadorn's supervision was brought to light. Diverse stories which had remained unlinked suddenly converged. Concepts and paths of research were defined : for ex. Homeosis, Physiological Genetics, 'Body plans' Allometry, Homologies of Process, Evolution as 'bricolage' and finally a critical essay on different perspectives on development.

These communications are by no means a 'bilan'. They are directions chosen by colleagues who liked the spirit of the symposium and promised themselves to repeat such workshops.

We were saddened by the sudden demise of Pr Pierre Tardent of the University of Zürich. We wish to thank Pr H. Tobler who read the communication which his colleague had prepared and Pr R. Stidwill for the obituary. We also wish to express our gratitude to Pr R. Nöthiger who agreed to review the communication for its publication.

MULLER ON DEVELOPMENTAL GENETICS

Raphael FALK

Received wisdom has it that Morgan, after his conversion to Mendelism in 1910, turned his back to embryology. Actually, Morgan's experimental work on Mendelian genetics should be viewed substantially as an inherent refinement of his theoretical conception of embryogenesis and development. Yet, Kohler (1994, *The Lords of the Fly*) suggests that although members of the fly group " did the most sustained and ingenious work in trying to reconnect Drosophila genetics with development... it was mainly the difficulty of inventing experimental systems that could compete with mainstream practices " that impeded efforts to reconnect genetics with development. According to Kohler the trouble was that the improvised routine of co-opting for developmental purposes the experimental systems of Drosophila, somehow kept turning experiments designed to illuminate development into experiments on segregation and crossing over.

My contention is that this interpretation of Kohler, even if correct for Morgan and some other researchers, does not apply to persons like Muller, Sturtevant, Curt Stern, and Beadle in collaboration with Tatum, to name a few, who knew how to formulate their developmental problems in terms that the genetic machinery could answer, rather than to superimpose developmental and physiological methods and question-posing on the genetic system.

As soon as Morgan comprehended Johannsen's distinction of genotype and phenotype he underscored the many-to-many relationships between genes and characters. Still, Morgan never viewed the genes as the " atoms of heredity " as Muller did. For Morgan the genes were only one of the elements that participated in cellular physiology. Morgan conceived of the genes as undergoing structural changes, just like any other cytoplasmic components, and hoped that they could do this without losing their fundamental properties.

Viewing the genes from the cell's point of view, rather than viewing the cells from the gene's point of view, constrained Morgan's capacity to frame relevant embryological questions that the genetic system he so successfully developed could answer. Muller, on the other hand, conceived of the genes, on pure theoretical grounds, as constant unchangeable entities (except for rare mutations), the guardians of " genetic information ". He accepted genes as the " atoms of development " : It is true that there are many-to-many relationships between genes and characters, if we view the genes from the vantage point of the characters, but viewed from the genes' perspective, every gene has *one* effect or product that may contribute to many characters (and characters may be affected by the specific products of many genes). Once Muller accepted the major role of the gene in developmental mechanics his working program on mutagenesis was a direct consequence. Developmental genetics was a leitmotiv in his work that grew directly out of his conception of the *Entwicklungs-mechanik*.

Muller repeatedly pointed out that he considered himself primarily a student of E.B. Wilson and a follower of his teaching. The extent of this influence is made most explicit in his two obituary appraisals of Wilson in 1943 (*American Naturalist*, 77 : 5-37 & 142-172) and 1949 (*Genetics*, 34 : 1-9), which are to a large extent Muller's own autobiographical scientific confessions, and in his very last paper, prepared for the Mendel Centennial Symposium of the Genetics Society of America, which he was already too ill to attend in 1964 (*American Naturalist*, 100 : 493-517). According to Muller the main thrust of developmental mechanics was directed to the protoplasmic notion, but the notion of the gene as the basic unit of development, was already immanent to the teaching of the founders of modern material, experimental biology, like Wilson and Strasburger. As he saw it, Wilson's central objective was " the problem of how the individual lies determined within the germ ". It was this issue of the egg determinants that guided Wilson on his move from embryology *sensu lato* to its formulation in cellular terms *sensu stricto*, and eventually to its more profound sense, of the biology of the gene : Muller rejected not only Driesch's vitalism but also the embryologists' loose terminology, and adopted a terminology that could one day be expressed in physico-chemical interaction terms. It is here that Muller, the material reductionist, deviated from main-stream embryology. As stated by Gilbert *et al.* (*Developmental Biology*, 173 (1996), 357-372) : " The basic paradigm of embryology, the idea that gave it structure and coherence, *was the morphogenetic field* ". For Muller any notion of a " field " not expressed in solid physico-chemical terms was an anathema that smacked of " mysticism ". Muller argued that nuclear-cytoplasmic interactions, as revealed in the fertilized egg cell, were the key for developmental mechanics free of " mysticism ". Muller's approach to the problem of developmental genetics may be divided into two aspects, the principal one and the practical one. In principle, embryogenesis should be reduced to the genic level, to be upheld by sound physicochemical conceptions. On the practical aspect, he preached for the need of detailed information on the phenotypic effects of as many genes and their mutations as possible and their interaction in the development and differentiation of the cell and the organism as a whole. The tools for such a phaenogenetic analysis were gene mutations. But although he prescribed how to achieve this end — in a scheme amazingly similar to that of the proponents of the knockout and foot printing genetics of today he had to admit that most of this was not yet obtainable at his time. Although Muller tried to elucidate general principles of genes' action, he constantly reminded us that mutagenesis provides a large array of possibilities to discern in detail specific developmental and physiological patterns : " Potentially by far the most analytical use to which the production of genetic changes by radiation may be put in studies of developmental processes is through its provision of mutant genes, the effects of which on development are then traced in detail. …The field is a virtually unlimited one since, theoretically, the method could be applied for each of the thousands of different genes capable of mutating, and even for each of the different mutant alleles of these genes ". And quoting an earlier statement by himself, he states : " …each gene must be considered as producing its own specific chemical material in the cell, …It is a task for the future to determine the composition of all these substances and the nature of the complicated interactions whereby they co-operate to make the organism what it is ". (Muller, 1954, in A. Hollaender (ed.), *Radiation Biology*, vol. 1. p. 459). Here Muller summarized in a nutshell his view of the genes as the entities providing the substructure that governs and regulates all metabolic and developmental pathways of living organisms, on firm physico-chemical principles. But rather than being part of these pathways, genes are on a level of organisation underneath them. This is the reason why the classic methods of physiology and embryology would not do in elucidating " gene function " as such.

Les apports de la génétique russe pré-stalinienne à la connaissance des mutants homéotiques de la Drosophile

Stéphane SCHMITT

Le concept d'homéosis suscite depuis une vingtaine d'années un grand intérêt en biologie. Les progrès de la génétique du développement depuis la fin des années 1970 sont ainsi liés pour une large part aux études portant sur les mutations homéotiques de la Drosophile. Par ailleurs, l'homéosis a également fourni aux évolutionnistes certains de leurs modèles les plus heuristiques. Ce succès spectaculaire n'a pas manqué de donner lieu à un retour en arrière, vers les origines de ce concept et les usages qui en ont été faits au cours de l'histoire de la biologie[1]. Malgré cela, de nombreux points restent encore à éclaircir, notamment en ce qui concerne la période allant de la création du terme aux théories de Richard Goldschmidt dans les années 1940. Nous allons nous intéresser ici aux travaux des prédécesseurs de cet auteur, en mettant l'accent sur le rôle, apparemment très important, de l'école russe de génétique dans les années 1920.

C'est le Britannique William Bateson qui est à l'origine du concept d'homéosis[2]. Son objectif était alors de recenser toutes les variations discontinues connues alors dans le règne animal ; il avait en effet la double conviction que seule la connaissance des mécanismes exacts de la variation pourrait conduire à la compréhension des modalités de l'évolution, et que, d'autre part, seules les variations discontinues jouent un rôle significatif dans l'évolution. Dans sa classification des différents types de variations discontinues, Bateson introduit la notion d'homéosis, remplacement d'un membre d'une série répétée de parties du corps (une série " méristique ") par un autre membre de la même

---

1. Edward B. Lewis, " Homeosis, the first 100 years ", *Trends in Genetics*, 10, 341-343.
2. William Bateson, *Materials for the Study of Variation*, London, Macmillan, 1894.

série. Ce type de transformation était connu en botanique sous le nom de " métamorphose "[3], terme jugé trop ambigu par Bateson.

L'embryologie expérimentale s'empare du concept au début du XXe siècle. Hans Przibram travaille alors, à Vienne, sur des cas de régénération atypiques (hétéromorphoses). Une telle étude a été entreprise quelques années auparavant par Curt Herbst chez des Crustacés, ce qui lui a permis de toucher du doigt le concept d'induction[4]. Mais Przibram réalise un travail plus synthétique sur les différents cas d'homéosis chez les Arthropodes[5]. Il interprète la plupart d'entre eux comme des " expériences de la nature ", et s'intéresse surtout aux causes mécaniques qui ont pu en être à l'origine. Le caractère éventuellement héréditaire de ces variations est à peine évoqué.

C'est en 1923 que paraît aux Etats-Unis le premier catalogue de mutants de la Drosophile où il est fait mention de cas d'homéosis (sans que ce terme y soit employé)[6] : les mutations *bithorax* et *bithorax-b*, qui provoquent chacune la transformation du troisième segment thoracique des mouches (qui porte normalement des balanciers) en une structure rappelant le second segment thoracique (qui porte normalement les ailes). Mais les auteurs ne s'intéressent guère à l'apport éventuel de ces observations à la biologie du développement. Conformément aux recommandations que Thomas Morgan formulera un peu plus tard, ils séparent nettement l'étude de la transmission des caractères et celle de leur expression[7].

La Russie des années 1920 est l'un des pôles les plus florissants de l'époque en génétique, bien que cette discipline n'y ait pénétré qu'assez tardivement, après la Révolution d'Octobre[8]. De nombreux instituts lui sont alors consacrés, dont deux à Leningrad et un à Moscou. Plusieurs facteurs expliquent ce succès, mais l'un des plus importants semble avoir été le prestige de la tradition d'his-

3. Ce terme avait été introduit par Goethe et repris par le botaniste anglais Masters. Voir Johann Wolfgang Goethe, " Versuch die Metamorphose der Pflanzen zu erklaeren ", dans *Schriften zur Morphologie*, Franckfurt-am-Main, Deutscher Klassiker Verlag, 1987 ; T.H. Masters, *Vegetable Teratology*, London, Hardwicke, 1869.

4. Curt Herbst, " Über die Regeneration von antennenähnlichen Organen an Stelle von Augen ", *Archiv für Entwicklungsmechanik der Organismen*, 2 (1896), 544-558. Sur Curt Herbst et son apport à l'*Entwicklungsmechanik*, consulter Jane M. Oppenheimer, " Curt Herbst contributions to the concept of embryonic induction ", dans Scott F. Gilbert (ed.), *A Conceptual History of Modern Embryology*, New York, Plenum, 1991.

5. Hans Przibram, " Experimentelle Studien ueber Regeneration ", *Wilhelm Roux' Archiv für Entwicklungsmechanik der Organismen*, 11 (1901), 321 ; " Die Homoeosis bei Arthropoden ", *Wilhelm Roux' Archiv für Entwicklungsmechanik der Organismen*, 29 (1910), 424-447.

6. Calvin B. Bridges et Thomas H. Morgan, " The Third Chromosome Group of Mutant Characters in *Drosophila melanogaster* ", *Carnegie Institute of Washington Publications,* 327 (1923).

7. Thomas H. Morgan, *The Theory of the Gene,* New Haven, Yale University Press. Sur cette séparation entre la génétique et l'embryologie, voir Garland E. Allen, " T.H. Morgan and the Split between Embryology and Genetics ", dans T.J. Horder, I.A. Witkowsky et C.C. Wylie (eds), *A History of Embryology*, New York, Cambridge University Press, 1987.

8. Sur la génétique russe à cette époque, voir A.E. Gaissinovitch, " The Origins of Soviet Genetics and the Struggle with Lamarckism, 1922-1929 ", *Journal of the History of Biology*, 13 (1980), 1-51.

toire naturelle héritée de l'époque tsariste, corrélé à la faveur dont a joui très tôt la théorie de la sélection naturelle, acceptée en Russie beaucoup plus largement que dans les pays occidentaux. Contrairement à ce qui s'est produit dans d'autres pays, il n'y a eu aucune rupture entre la morphologie classique et les approches plus " physiologiques " telles que l'embryologie expérimentale.

On considère que la génétique des populations est née dans ce contexte favorable, en particulier grâce aux travaux de Sergei Tchetverikov, de l'institut Koltsov de Biologie Expérimentale de Moscou. Il y effectue de nombreuses études sur les populations sauvages de *Drosophila melanogaster*. C'est à cette occasion que deux de ses élèves, Boris L'vovitch Astaurov et E.I. Balkaschina, découvrent deux nouveaux mutants homéotiques.

Le premier d'entre eux est un mutant récessif qui se manifeste par la présence de structures alaires à la place des balanciers, phénotype très similaire à ceux de *bithorax* et de *bithorax-b*. Il est baptisé *tetraptera* par Astaurov et présenté dans une revue occidentale en 1929[9]. C'est à la suite de cet article que Balkaschina publie son étude de la mutation récessive " *aristopedia* " (sic), concernant cette fois la partie distale de l'antenne, transformée en tarse[10]. Ces deux auteurs considèrent explicitement les phénotypes observés comme relevant de l'homéosis, et se réfèrent à Bateson et à Przibram. Pour la première fois, le terme " homéosis " est donc appliqué à un mutant de la Drosophile : il s'agit par conséquent d'une étape essentielle vers la notion de " gène homéotique ".

En outre, Balkaschina fait le lien entre ce type d'homéosis, héréditaire, et les travaux sur la régénération : " Cette série de changements homéotiques, qui correspondent à des changements dans la substance héréditaire et sont réalisés dans l'organisme durant sa période embryonnaire, est parallèle à une série de transformations homéotiques analogues qui ont lieu par régénération, dans ce dernier cas, à la place de l'organe extirpé, croît un régénérant hétéromorphe […]. D'après l'observation du parallélisme entre les hétéromorphoses héréditaires et régénératrices, il s'ensuit que les forces qui provoquent un changement dans le développement d'un organe donné agissent de la même façon dans l'organisme qu'il s'agisse d'actions extérieures (c'est-à-dire l'extirpation de l'organe) ou intérieures (c'est-à-dire par variabilité génotypique) ".

Cet intérêt pour les " forces " (*Kräfte*) à l'origine de l'homéosis traduit une influence de l'*Entwicklungsmechanik* et de la génétique allemande. En effet, il existe à cette époque en Allemagne une volonté d'expliquer les relations entre génotype et phénotype dans le cadre d'une " génétique physiologique ". Ces

---

9. Boris L. Astaurov, " Studien über die erbliche Veränderungen der Halteren bei *Drosophila melanogaster* Schin ", *Wilhelm Roux' Archiv für Entwicklungsmechanik der Organismen*, 115 (1929), 424-447.

10. E. I. Balkaschina, " Ein Fall der Erbhomoeosis (die Genovariation " *aristopedia* ") bei *Drosophila melanogaster* ", *Wilhelm Roux' Archiv für Entwicklungsmechanik der Organismen*, 115 (1929), 448-463.

préoccupations ont trouvé en Russie un terrain des plus favorables et s'y manifestent notamment par un intérêt pour la chronologie de l'action des gènes sauvages et de leurs mutants au cours du développement.

Balkaschina compare ainsi minutieusement le développement des disques imaginaux chez des Drosophiles sauvages et *tetraptera*, pour en déduire à quel moment agit le gène touché par la mutation. Des coupes sont réalisées en parallèle dans les disques imaginaux des antennes d'individus mutants et sauvages, à des moments précis des stades larvaire et pupal. De ces observations, Balkaschina déduit que " l'action du gène *aristopedia* se manifeste au cours du développement embryonnaire de la mouche en stimulant la segmentation des disques imaginaux des antennes, à un stade précoce du développement, à savoir au début de la segmentation des pattes, et par l'orientation de la différenciation vers un développement de tarse et non d'antenne ". Elle propose donc une interprétation originale de la mutation *aristopedia*, et par conséquent du mode d'action de l'allèle sauvage correspondant. Selon ses observations, la mutation provoquerait une altération dans la chronologie du développement des disques imaginaux des antennes, qui est accéléré au point d'avoir lieu en même temps que celui des disques imaginaux des pattes, dont la différenciation lui est normalement très antérieure. Cette simultanéité conduit les disques antennaires à se développer en forme de pattes. Cette interprétation sera reprise et étoffée par Richard Goldschmidt, qui se référera alors explicitement aux travaux de Balkaschina.

Les Russes sont donc les premiers à étudier des mutants homéotiques sous l'angle de la " génétique physiologique " et du développement. Ils envisagent également les implications évolutives de ce phénomène. Ils suivent ainsi leurs prédécesseurs dans l'établissement de liens d'homologie entre les organes concernés par l'homéosis. Selon Astaurov, " on peut bien sûr affirmer, dans tous les cas d'homéosis, qu'un organe n'acquiert pas aléatoirement les caractéristiques d'un autre, la transformation n'est pas déterminée par la simple proximité spatiale mais elle trouve son origine dans la profonde similitude [*tieferen Ähnlichkeit*] entre les deux organes. On se représente mal qu'une patte d'insecte puisse se transformer en aile ou qu'à la place d'un oeil de Vertébré apparaisse un membre, et la raison n'en est pas tant qu'un tel fait n'est pas connu, mais qu'aucune donnée d'embryologie ou d'anatomie comparée ne révèle une quelconque ressemblance dans le développement ontogénétique et phylogénétique de ces organes [...]. Etant donné que le cas de *tr*, à mon avis, représente un cas typique d'homéosis, nous pouvons affirmer avec un fort taux de probabilité, sans en appeler pour cela aux observations de la morphologie comparée, que la transformation des balanciers en ailes n'est pas un caprice aléatoire de la variabilité génétique mais qu'elle est liée à l'affinité intime de ces organes et qu'elle prouve que les ailes et les balanciers sont des organes homodynamique ".

Balkaschina conclut de même à propos d'*aristopedia* que " l'apparition d'une telle transformation chez *Drosophila melanogaster* offre la preuve que même chez des organismes hautement différenciés tels que les Diptères, les structures embryonnaires méristiques conservent la capacité de modifier la direction prise par leur développement et de se changer en des structures homologues. La mutation *aristopedia* est une preuve de l'isopotentialité des disques imaginaux des Diptères ."

Par ailleurs, Astaurov considère la mutation *tetraptera* comme un atavisme et l'interprète comme la résurgence de facteurs embryonnaires ancestraux, " la réapparition de relations causales [*kausale Zusamenhänge*] perdues depuis longtemps dans leur état original, c'est-à-dire [comme] une réversibilité (partielle naturellement) du processus évolutif ".

La parenté de cette interprétation avec celles qu'avançaient les auteurs allemands et autrichiens à propos de l'hétéromorphose est frappante et ne laisse subsister aucun doute sur l'influence de l'*Entwicklungsmechanik* sur la biologie russe pré-stalinienne, y compris la génétique. Cet emprunt n'a néanmoins rien de docile et s'inscrit dans une tendance générale des chercheurs russes à ne négliger aucun des différents types d'approches d'un phénomène donné. À cet égard, citons l'appel d'Eugen Schultz, en 1902, à unir " les méthodes historique [c'est-à-dire phylogénétique] et expérimentale " dans l'étude de la régénération[11].

Balkaschina aborde enfin au sujet d'*aristopedia* le fait que les mutations géniques peuvent affecter non seulement des caractères d'espèce, mais aussi des caractères définissant des rangs taxinomiques supérieurs tels que l'unique paire d'ailes des Diptères ; cette idée sera reprise et développée par Bridges et Dobzhansky et surtout par Goldschmidt.

Les généticiens russes ont donc joué un rôle déterminant dans l'histoire de l'homéosis. En effet, ce sont eux qui, les premiers, appliquent ce concept à des mutants de la Drosophile, et ce n'est qu'avec l'arrivée aux Etats-Unis d'un membre de l'école russe, Theodosius Dobzhansky, que les Américains l'adopteront.

L'approche des Russes est en outre synthétique : ils associent à l'analyse génétique une étude morphologique et anatomique détaillée, ils envisagent l'intérêt des mutations pour la compréhension du développement et attribuent une place importante aux considérations phylogénétiques dans leur interprétation.

Si leurs articles dénotent un certain nombre d'influences externes (notamment l'*Entwicklungsmechanik*, à laquelle Astaurov " emprunte " l'explication de l'atavisme et Balkaschina la classification, établie par Przibram des diffé-

---

11. Eugen Schultz, " Über das Verhältnis der Regeneration zur Embryonalentwicklung und Knospung ", *Biologisches Zentralblatt,* 22 (1902), 360-368.

rents cas d'homéosis), ils traduisent une grande ouverture d'esprit et une pensée originale, et il est indéniable qu'ils ont, en retour, inspiré de nombreuses réflexions sur l'homéosis, à commencer par celle de Goldschmidt[12].

Après 1930, les travaux du groupe de Moscou sont interrompus pour des raisons politiques. Cependant les idées de l'école russe ne sont pas totalement perdues, et l'exil de certains de ses membres contribue au contraire à les disséminer. C'est ainsi que Theodosius Dobzhansky, émigré aux Etats-Unis (après avoir travaillé à Leningrad), va être l'un des principaux contributeurs à l'élaboration de la théorie synthétique de l'évolution[13].

Ce n'est sûrement pas un hasard si le titre de l'article de Bridges et Dobzhansky, publié en 1933, reprend mot pour mot celui de l'article de Balkaschina de 1929[14]. On y retrouve la tradition russe consistant à mêler analyse génétique et morphologique. La nouvelle mutation, découverte par Bridges en 1931 et baptisée *proboscipedia* est présentée comme un cas d'homéosis. Il s'agit en effet " d'une forte altération de la structure de la région buccale, transformant les lobes oraux en appendices ayant une forme d'antenne ou de tarse ".

La plus grande partie de l'article est consacrée à une description morphologique de *proboscipedia* et à sa comparaison avec le type sauvage. Les auteurs concluent finalement que la mutation rend les pièces buccales de la Drosophile semblables à celles d'insectes appartenant à des ordres jugés moins évolués. Ils soulignent qu'" une unique mutation génique est capable de changer des caractères auxquels les taxinomistes accordent une signification considérable ". Cette remarque s'adresse à ceux qui, en France et en Allemagne, sont plutôt favorables à la prédominance d'une hérédité cytoplasmique, alors que le noyau et les gènes ne contrôleraient que des caractères de moindre importance, tels que la couleur des yeux. Les mutants homéotiques infirment cette opinion, et Bridges et Dobzhansky n'hésitent pas à insister sur ce point dans un article destiné à paraître dans l'une des plus prestigieuses revues allemandes.

Cependant, ils ajoutent qu'" il est évident, bien sûr, que des mutations comme *bithorax* et *proboscipedia* ne présentent pas l'apparence de nouvelles espèces et encore moins de nouvelles familles ou de nouveaux ordres. *Proboscipedia* et *bithorax* doivent être considérées comme des formes de l'espèce *Drosophila melanogaster*. Il est suffisamment bien connu que des espèces ou des genres diffèrent entre eux par de nombreux gènes. Ainsi, les différences spécifiques et génériques ne peuvent pas survenir par une seule mutation ponc-

---

12. Richard B. Goldschmidt, *The Material Basis of Evolution*, New Haven, Yale University Press, 1940.

13. Voir M.B. Adams (ed.), *The Evolution of Theodosius Dobzhansky. Essay on His Life and Thought in Russia and America*, Princeton, Princeton University Press, 1994.

14. Calvin B. Bridges et Theodosius Dobzhansky, " The mutant 'proboscipedia' in *Drosophila melanogaster*. A case of hereditary homoeosis ", *Wilhelm Roux' Archiv für Entwicklungsmechanik der Organismen*, 127 (1933), 575-590.

tuelle. Il est toutefois intéressant de savoir que même les structures les plus " fondamentales " peuvent être " fondamentalement modifiées par une unique mutation génique ".

La structure des pièces buccales, l'absence d'une seconde paire d'ailes font partie des caractères permettant de définir des rangs taxonomiques élevés. Elles peuvent ainsi être affectées par un événement ponctuel, et ces modifications sont transmissibles à la descendance. Il serait aisé de rattacher ces faits à une conception saltatoire de l'évolution, celle de Bateson par exemple, et c'est ce que fera Goldschmidt quelques années plus tard. Mais Dobzhansky est en train de participer à la construction de la théorie synthétique de l'évolution, néo-darwinienne, et dont l'un des piliers est le caractère continu et progressif de la spéciation. Il ne peut donc adhérer à une telle hypothèse, et s'empresse de couper court à toute conclusion qui irait dans ce sens.

Il n'en demeure pas moins que cet article marque une étape importante dans l'histoire de l'homéosis. En effet, sous l'influence de la *diaspora* des biologistes russes, ce concept est adopté par la génétique américaine, qui jusqu'alors l'avait totalement ignoré, et il est indéniable que cette influence a joué un grand rôle dans les travaux ultérieurs, notamment ceux de Richard Goldschmidt, qui utilisera l'homéosis dans sa tentative pour d'élaboration d'une synthèse entre génétique, embryologie et évolution.

REMERCIEMENTS

Je tiens à remercier chaleureusement Nadine Peyrieras pour sa patience et sa disponibilité lors de la rédaction du mémoire de DEA dont s'inspire cet article. Tous mes remerciements également à Charles Galperin qui m'a permis d'assister à ce congrès, ainsi qu'à Michel Morange, Hervé Le Guyader et Claude Debru pour leur soutien et leurs conseils.

L'HOMOEOSIS : DU PHÉNOTYPE MUTANT A LA FONCTION DÉVELOPPEMENTALE DU GÈNE

Nadine PEYRIERAS

Dans le contexte contemporain de la Biologie du Développement, l'homoeosis est liée à la fois aux questions évolutives et embryologiques et son étude se poursuit avec les outils de la génétique formelle et de la génétique moléculaire. Son histoire touche différentes disciplines et révèle les questions et les stratégies expérimentales qui leurs sont propres.

Homoeosis est le terme choisi par William Bateson en 1894 pour décrire certains types de variation[1].

It is desirable and indeed necessary that such Variations, which consist in the assumption by one member of a Meristic series, of the form or characters proper to other members of the series, should be recognized as constituting a distinct group of phenomenon (p. 84).

Le terme Homoeosis permet de souligner une propriété essentielle des variations observées : ...*the essential phenomenon is not that there has merely been a change, but that something has been changed into the likeliness of something else* (p. 85).

William Bateson travaille avec des hypothèses fortes qu'il n'a pas les moyens d'explorer. L'évolution peut-elle procéder par le biais de telles variations et l'homoeosis peut-elle rendre compte de différences inter-spécifiques ? Malgré l'impossibilité d'aborder ces questions, William Bateson élabore les fondements descriptifs et conceptuels d'une variabilité de type homoeotique au sein du règne animal. Le contexte de son étude est celui de l'évolution et de la spéciation. Sa critique de la méthode embryologique pour résoudre ces questions l'éloigne d'un lien entre l'homoeosis et le développement embryonnaire.

Au XXe siècle, l'homoeosis est attachée au contexte de l'évolution ainsi qu'à celui de l'embryologie expérimentale. Ce rapprochement est le fait de la

---

1. William Bateson, *Materials for the Study of variation*, London, Macmillan and Co., 1894.

tradition allemande de l'*Entwicklungsmechanik*[2]. En particulier, de nombreux travaux traitant de la régénération chez les Arthropodes ont permis la description de diverses anomalies morphologiques relevant de l'homoeosis. Curt Herbst s'est attaché à l'aspect causal de la régénération. Eugen Schultz privilégie une description du caractère atavique des organes régénérés. Chronologiquement, ces travaux s'insèrent entre l'oeuvre de William Bateson et les premières descriptions de cas héréditaires d'homoeosis. Ils imposent une interprétation des observations de William Bateson dont lui même s'était gardé et qui sera en partie utilisée par les généticiens.

Deux écoles vont contribuer aux premières descriptions de cas héréditaires d'homoeosis. Les collaborateurs de Thomas H. Morgan privilégient l'étude de la transmission héréditaire des caractères et la cartographie génétique[3]. Ils ne se réfèrent à l'homeosis que sous l'influence de l'école russe de génétique par l'intermédiaire de Théodore Dobshansky[4]. Dès le début des années 1920, les généticiens russes réalisent une forme de synthèse des descriptions de l'embryologie expérimentale et d'une approche génétique de l'homoeosis. Cela est illustré par la publication de E.I. Balkaschina[5] avec son souci de mettre en évidence les conséquences phénotypiques les plus précoces de la mutation *aristopedia* chez *Drosophila melanogaster*. Elle est amenée à étudier les phénotypes larvaires et sa description morphologique s'appuie sur l'observation de coupes histologiques. Elle se réfère aux travaux de William Bateson, Curt Herbst et Eugen Schultz. Tout en soulignant le caractère héréditaire de la transformation observée, la principale conclusion de son étude est celle d'une relation embryologique, qualifiée d'équipotentialité, entre les ébauches d'antenne et de patte.

Calvin B. Bridges et Théodore Dobshansky publient en 1932 la description du mutant *proboscipedia*. Les termes de leur étude nous rapprochent du travail de E.I. Balkaschina qui est cité. Ils ne font par contre pas référence aux auteurs qui l'ont précédée. Leurs descriptions se limitent au phénotype de l'adulte. Leur discussion nous permet de percevoir le contexte d'interprétation des phénomènes observés. *Proboscipedia* est rapproché des autres cas héréditaires d'homoeosis : *bithorax*, *bithorax-b* et *tetraptera*.

A single gene-mutation is, therefore, able to change characters of the kind to which taxonomists ascribe considerable significance.

2. Stéphane Schmitt, *L'histoire de l'homéosis des origines à Richard Goldschmidt*, DEA, Paris, Université Paris 7, 1996.

3. Calvin B. Bridges and Thomas H. Morgan, *The Third group of Mutant Characters in Drosophila melanogaster*, Washington,Carnegie Institute, 1923 .

4. Calvin B. Bridges and Theodore Dobzhansky, " The Mutant " proboscipedia " in Drosophila melanogaster — A case of hereditary homoösis ", *Wilhelm Roux' Archiv für Entwicklungsmechanik der organismens*, 127, (1933), 575-590.

5. E.I. Balkashina, " Ein Fall des Erbhomoösis (die Genovariation " aristopedia ") bei Drosophila melanogaster ", *Wilhelm Roux' Archiv für Entwicklungsmechanik der organismens*, 115, (1929), 448-463.

This fact seems to be contradictory to the view repeatedly expressed by certain authors, according to which mutation changes affect only " superficial" structures, of the kind distinguishing different varieties of the same species or, as a maximum, different species of the same genus. These authors elaborate this idea further, to prove that only characteristics of the varieties, species, and, perhaps, genera are determined by the action of the genes, and are, therefore, capable of being changed by point-mutations. The "fundamental" characters, such as those distinguishing families, orders, and classes, are supposed to be determined not by genes, but by some " central" part of the germplasm, not divisible into genes, and associated with the cytoplasm.

Its is obvious, of course, that mutations similar to bithorax and proboscipedia do not represent appearances of new species, and still less of new families and orders. Proboscipedia and bithorax must be classified as forms of the species Drosophila melanogaster. It is sufficiently well known, that different species and genera differ from each other in many genes. Hence the specific and generic differences can not possibly arise by a single point-mutation. It is interesting, however, to know that even " fundamental" structures can be " fundamentally" changed by a single gene-mutation (p. 589).

Le résumé de la publication terminant l'article mentionne la ressemblance du mutant avec les insectes inférieurs, référence à l'atavisme, et l'homologie des appendices de la tête entre eux et avec les pattes. L'accent est donc mis sur les concepts de l'évolution et de l'embryologie expérimentale. La suggestion d'un rôle des gènes dans l'élaboration des caractères ayant une importante " valeur taxonomique " reste éloignée de leur rôle dans le développement embryonnaire.

L'étude des cas héréditaires d'homoeosis abordée par les écoles russes et américaines de génétique n'est pas guidée par un questionnement sur le rôle des gènes dans le développement embryonnaire. Par contre, Richard B. Goldschmidt au début des années 1930 et Conrad H. Waddington un peu plus tard ont eu explicitement cette préoccupation dans des approches expérimentales de l'homoeosis. Cependant, leurs choix conceptuels et expérimentaux ne permettront pas d'avancée décisive. De plus, ils vont tous les deux s'opposer à Edward B. Lewis dont l'approche expérimentale est la seule réellement heuristique.

Conrad H. Waddington écrit en 1972 dans " The Morphogenesis of Patterns in *Drosophila* "[6] : *I believe it is worthwhile recalling this earlier work, because Drosophila still remains the most favourable system for investigation of questions concerning the organization of genes activities.*

The first studies which attempted to trace the development of mutant phenotypes in Drosophila were made by Goldschmidt (1933, 1935, 1937). He

---

6. C.H. Waddington, *The Morphogenesis of Patterns in Drosophila*, in *Developmental systems in insects*, Counce (eds), (Londres, S.J. et Waddington, C.H. Academic Press, 1972), Tome II.

approached the matter from the point of view of a geneticist and asked : " can I, by observing the timing and nature of the developmental modifications produced by the gene deduce something about the nature of gene's primary action ? " In practice, he was not very successful in doing this. The mutant phenotypes he studied (i.e. abnormal wing shapes in Drosophila, pigmentation patterns and sexual dimorphism in Lymantria) were too complex to be expressed in chemical terms (p. 499).

R.B. Goldschmidt a en particulier étudié le mutant homoeotique *podoptera* présentant une très faible pénétrance, une grande variabilité et impliquant plusieurs loci. Ce choix, justifié par ses préoccupations conceptuelles, ne lui a pas permis d'approcher la logique de fonctionnement des gènes homéotiques et l'a aussi éloigné de l'hypothèse " un gène, une enzyme " que G.W. Beadle, B. Ephrussi et Haldane ont pu formuler en étudiant des systèmes plus simples.

C.H. Waddington souligne sa propre contribution : *The question posed was not " What does this particular gene do ? " but rather " How is the development of this organ affected by the genotype of the cells composing it ?...*

Drosophila provided the opportunity to demonstrate that a " developmen-tal pathway " is a process which is controlled at every point by the action of the genes (Waddington, 1939, 1940) (p. 500).

Cependant la compréhension du rôle développemental des gènes homéotiques n'a pas résulté de ce type de questionnement. En effet, les plus remarquables de leurs propriétés sont liées à leur arrangement chromosomique et à ce qu'il suggère de leur origine évolutive et de leur mode de fonctionnement. En 1962, C.H. Waddington rejette cette logique[7]. Non seulement il dénonce comme une imposture l'approche expérimentale de E.B. Lewis mais il trahit aussi ses observations. E.B. Lewis aborde le mode de fonctionnement des gènes homoeotiques à partir de l'étude du pseudoallélisme. Cette approche a été encouragée par les observations de Calvin B. Bridges en 1935[8]. Ce dernier écrit dans *The Journal of Heredity* : *In my first report on duplications at the 1918 meeting of the A.A.A.S. , I emphasized the point that the main interest in duplications lay in their offering a method for evolutionary increase in lengths of chromosomes with identical genes which could subsequently mutate separately and diversify their effects* (p. 64).

E.B. Lewis établit comment la duplication des gènes au sein du complexe *Bithorax* de gènes homéotiques permet d'expliquer la morphologie des segments du corps de l'insecte[9]. Décrire le mode de fonctionnement du complexe *Bithorax* n'est pas une approche pertinente pour C.H. Waddington. Dire

---

7. C.H. Waddington, *New Patterns in Genetics and development*, Columbia, University Press, 1962.

8. Calvin B. Bridges, " Salivary chromosome maps ", *J. Heredity*, 26, (1935), 60-64.

9. E.B. Lewis, " A gene complex controlling segmentation in Drosophila ", *Nature*, 276, (1978), 565-570.

qu'une mutation est responsable du développement d'une drosophile portant deux paires d'ailes par la transformation d'un segment thoracique est un niveau d'explication dont il ne peut pas se satisfaire. Cette observation n'est pour lui qu'une source de questions.

Depuis C.H. Waddington, la question du lien entre le phénotype mutant et la fonction développementale du gène est clairement formulée mais résiste toujours à une approche expérimentale directe. La manière dont le gène homoeotique réalise sa fonction reste énigmatique. Cependant, la compréhension de la logique de fonctionnement des gènes homoeotiques qui a occupé E.B. Lewis depuis 1950 permet la description d'une hiérarchie de fonction entre les gènes illustrée par les modèles de E.B. Lewis[10] et A. Garcia Bellido[11]. Le succès de ces travaux est lié à la mise en évidence, à partir de 1984, de la conservation évolutive au sein du règne animal du complexe *Bithorax* et de certaines de ses propriétés.

10. *Idem.*

11. A. Garcia-Bellido, " Homoeotic and atavic mutations in insects ", *American Zoologist*, 17, (1977), 613-629.

PIERRE TARDENT†

Robert STIDWILL

Pierre Tardent who passed away on Sunday, June 8th, 1997 at the age of 70 was a passionate biologist. Born on May 21, 1927 in the small provincial town of Langenthal in the Swiss Canton of Berne he attended the higher schools in Burgdorf and then enrolled as a biology student at the University of Berne. At the age of 26 he had earned his Ph.D. by completing a doctoral thesis supervised by his mentor Fritz Lehmann. It was through the influence of the eminent developmental biologist Fritz Baltzer that Pierre Tardent received his first postdoctoral position. He was appointed as an assistant at the famous " Stazione Zoologica di Napoli " which goes back to the times of Ernst Haeckel and its founder Anton Dohrn. This first position proved to be formative for Pierre Tardent's entire professional career. In Naples where he ended up staying for ten years, Pierre Tardent interacted with distinguished visiting scientists from all over the world, establishing life long contacts and friendships. But it was also here at the shores of the " Golfo di Napoli " where he discovered his love for marine biology and acquired his profound knowledge of the marine fauna. In 1963, Ernst Hadorn, then Director of the Zoological Institute at the University of Zürich, appointed Pierre Tardent as assistant professor. Four years later, Pierre Tardent became an associate professor and, in 1968, a full professor. His research in Zürich focused on the regenerative capacities and other developmental aspects of coelenterate biology. He quickly grasped and applied the techniques of the young and ever since thriving field of cell biology. Pierre Tardent, however, will be best remembered by generations of biology and medical students as an enthusiastic and inspiring teacher. Both his lectures and laboratory courses were fascinating. They were only surpassed by his excursions in the field. Be it on one of his numerous visits to the French marine biological station in Banyuls-sur-mer or on a bird-watching excursion early in the morning somewhere around Zürich — this was where he really could live out his passion — biology. His broad interest in and knowledge about biological diversity ranging from invertebrates to reptiles, birds and mammals allowed him to motivate, stimulate and inspire any audience. With Pierre Tardent's passing, we lost a true biologist.

Pierre TARDENT
in his office at the Zoological Institute, University of Zurich

ERNST HADORN (1902-1976) : A PIONEER OF MODERN DEVELOPMENTAL GENETICS

Pierre TARDENT†

Shortened version of a lecture prepared for the XXth International Congress of History of Science in Liège, 20-26 July 1997.

After the sudden and unexpected death of Pierre Tardent on June 8th 1997, the lecture was delivered by Prof. Dr. H. Tobler.

Ernst Hadorn, born on May 31st in 1902, started his professional life as an elementary school teacher in a tiny Bernese village. From 1925 to 1930, he studied zoology at the University of Berne where his mentor was Fritz Baltzer (1884-1973), a student of Theodor Boveri (1862-1915). After his master's degree, Hadorn resumed teaching in a pre-University school. At home, he had installed a small laboratory where he performed all experiments for his PhD-thesis. In the course of the thesis, Fritz Baltzer became aware of the unusual qualities of Ernst Hadorn and arranged for him a postdoctoral year which was sponsored by the Rockefeller Foundation. Already in 1939, Hadorn became professor of Zoology at the University of Zurich, and in 1943, he was appointed director of the Zoological Institute which he headed until his retirement in 1972. Only four years later, he died on April 4, 1976.

Ernst Hadorn was an exceptional personality and scientist. Almost instinctively he " smelled " where the essential problems in developmental biology were. He then pursued them with perseverance and success. He was a reductionist who despised metaphysical thoughts. But he patiently discussed views even when they were not to his taste. He was fascinated by the complexity of developmental processes and anxious to know the mechanistic principles that govern them. His life-long concern was the search for the link between hereditary factors and ontogenetic processes. Hadorn's methods were simple and often unconventional. His credo was that normality can only be understood through the study of anomalies, either produced experimentally or offered by genetic mutations.

His work can be roughly divided into three phases : the first part, still following the research interests of Baltzer, focused on the functions of the cytoplasm and the nucleus and their interactions in the embryogenesis of amphibia.

For this purpose, Hadorn produced hybrid newt merogones by fertilizing enucleated eggs of *Triturus helveticus* with sperm of *Triturus cristatus*. Such animals underwent only limited development, showing that nucleus and cytoplasm have to match. This requirement, however, does not apply to all cells : By transplanting tissues and organs from lethal merogones into normal larvae, Hadorn demonstrated that certain cell types now survived and contributed to adult organs whereas others died cell-autonomously. After a few years, however, Hadorn became aware that working with amphibia had its limits because amphibian genetics was practically non-existent due to the lack of genetic mutations.

During his first sabbatical in 1936 at Rochester University (USA) where he collaborated with Curt Stern, he " discovered " *Drosophila* and immediately saw the great potential of this organism. Mutations that cause lethality at various stages of development must certainly identify important genes and developmental processes. From now on, Hadorn turned his attention to *Drosophila* and lethal mutations. His objectives remained the same : what are the factors that control developmental processes and lead to morphological and biochemical characters ? His thorough investigations of lethal factors in *Drosophila* revealed that these mutations played their fatal role at discrete phases of development and in specific organs. From these observations, he formulated his famous concept of " time- and tissue-specific action of genes ", later so beautifully verified by molecular biology.

In the 50s, Ernst Hadorn became captivated by the imaginal discs of *Drosophila*, their determination and differentiation. This work forms his third and last opus. Triggered by moulting hormones, each disc differentiates into a particular part of the fly, such as leg, thorax or genitalia, showing that the adult fly was mosaically built from a set of discrete disc primordia. By cutting discs into defined fragments and transplanting these into host larvae ready to pupate, Hadorn and his students showed that the mosaic character extends even to regions within a disc (Fig. 1).

His next sharp question was : is a particular state of disc determination stable, or is it susceptible to changes ? To solve this problem, Hadorn forced the cells of a disc to undergo an almost infinite number of divisions by repeatedly transplanting fragments of imaginal discs into adult female flies where the fragments grew, but did not differentiate. From time to time, a piece of the grown, and still growing, fragment was subjected to metamorphosis by transplantation into a host larva (Fig. 2).

In the first few transfers, the tested disc fragments faithfully reproduced their original state of determination. Later, however, structures typical of other discs appeared, indicating that transdetermination had occurred. These transdetermination events followed certain rules : their frequency positively correlated with the disc's intensity of proliferation ; and the direction of trans-determination showed a global orientation (Fig. 3, 4).

Hadorn was convinced that determination in imaginal discs is not a rigidly fixed, irreversible state, but the result of a dynamic steady-state, and that trans-determination is the consequence of a disequilibration of this steady-state, leading to the activation of a new set of genes. For each state of determination, he postulated a master gene which would direct " teams of genes " whose expression was required to construct a wing, a leg or an antenna. This was nothing less than a visionary, but correct prediction of what is known today about master genes controlling developmental processes. What he did not dare to hope is that such master genes could one day be isolated, manipulated, transferred and expressed at ectopic sites of an organism or in a foreign species.

Ernst Hadorn has made essential contributions to the understanding of the genetic control of developmental processes and has paved the way for modern developmental genetics. He was certainly one of the pioneers who brought together developmental biology and genetics. The impact of his work is documented by three seminal publications reflecting and summarizing the three phases of his research activities :

E. Hadorn, " Experimentelle Entwicklungsforschung, im besonderen an Amphibien ", Verständliche Wissenschaft, Vol. 77 (Berlin, Heidelberg, New York, Springer-Verlag, 1970).

E. Hadorn, *Developmental genetics and lethal factors*, London, Methuen & Co. ; New York, John Wiley & Sons, 1961.

E. Hadorn, " The Genetics and Biology of Drosophila ", in M. Ashburner & T.R.F. Wright (eds), *Transdetermination*, Vol. 2c, Academic Press, 1978, 555-617.

FIGURE LEGENDS

Ernst HADORN
at the microscope during his serial transplantation experiments
with imaginal discs

Ernst Hadorn in seinem Labor (ca. 1965) — Foto W. F. Böhm

Fig. 1

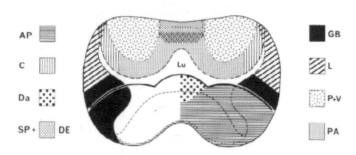

Mosaic character of the genital disc, revealed by subjecting defined fragments of the disc to immediate differentiation in a metamorphosing host (modified from Ehrensperger, Mitteil. Aarg. Naturf. Gesellschaft 30 : 145-237, 1983). The first fate map of a genital disc was constructed by Hadorn, Bertani & Gallera, Roux' Arch. Entw.-Mech. 144 : 31-70, 1949. Abbreviations : AP = anal plate, C = clasper, Da = hindgut, DE = ejaculatory duct, GB = genital arch, L = lateral plate, Lu = lumen, P = paragonia, PA = penis apparatus, Sp = sperm pump, V = vas deferens.

Fig. 2

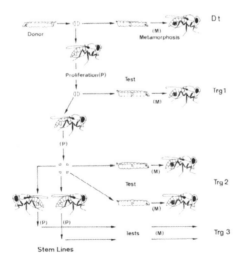

Method for long-term *in vivo* culture of blastemas from imaginal discs. Left side : stem line ; right side : differentiated test implants in metamorphosing hosts. Dt : Direct test ; Trg 1-3 : Transfer generations (from Hadorn, 1978).

Fig. 3

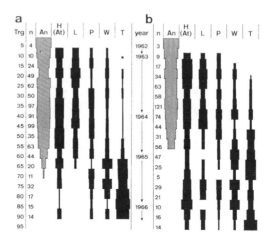

Performance of disc fragments after long and repeated culture *in vivo*. a, b : two cultures, each initiated with half a genital-anal disc. Vertical axis shows years during which the experiments were done, number of transfer generations (Trg) and of transferred disc fragments (n). Crosshatched area indicates structures of the original disc (An = analia). Black area shows transdetermination to H (At) = head (antenna), L = leg, P = palpus, W = wing, T = thorax.

Fig. 4

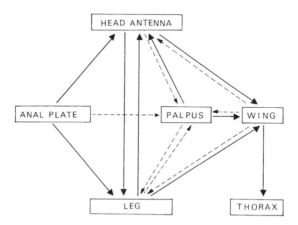

Direction of transdetermination shows a global orientation starting from analia (left) and leading towards wing and thorax (mesothoracic structures) as " end stations " (right).

DU LIGNAGE CELLULAIRE AU COMPARTIMENT. RECHERCHES SUR LA NAISSANCE DE LA GÉNÉTIQUE DU DÉVELOPPEMENT

Charles GALPERIN

J.E. Sulston et ses collaborateurs ouvraient leur grande étude sur le lignage cellulaire embryonnaire du nématode *Coenorhabditis elegans*, devenu un organisme modèle : *This report marks the completion of a project begun over one hundred years ago namely the determination of the entire cell lineage of a nematode*[1].

D'où vient ce projet qui va transformer les études embryologiques ?

L'ÉCOLE AMÉRICAINE DU LIGNAGE CELLULAIRE

Nous suivrons les travaux de l'école américaine en raison de la clarté de ses buts, de l'importance de ses travaux, des nombreuses discussions qui ont eu lieu à la station de biologie marine de Woods Hole. Cette école peut être illustrée en particulier par Edmund B. Wilson (1856-1939), le fondateur de la cytogénétique, l'ami de Thomas Hunt Morgan. Ses travaux sont parallèles à ceux de l'école allemande, ceux de Théodor Boveri en particulier[2]. Boveri, le dédicataire du grand livre de Wilson : *The cell in Development and Inheritance*, (1re éd. 1896). T.H. Morgan de même aura pour ami Hans Driesch.

Le choix de l'école américaine est évidemment dû à l'existence de la génétique de la Drosophile et les liens qui vont tenter de s'établir plus tard, et difficilement, entre les premiers travaux sur le lignage chez les invertébrés et l'analyse génétique d'un diptère dont les étapes du développement ne permettent précisément pas de suivre directement le lignage des cellules.

1. J.E. Sulston, E. Schierenberg, J.G. White and J.N. Thomson, " The Embryonic Cell-Lineage of the Nematode *Caenorhabditis elegans* ", *Dev. Biol.*, 100 (1983), 64-119.
2. Th. Boveri, " Uber die Entstehung des Gegensatzes zwischen den Geschlechtszellen und den somatischen Zellen bei *Ascaris Megalocephala* ", *Sitz. Ges. Morphol. Physiol.*, 8 (1893), 114-125.

L'école américaine a été fondée par Charles Otis Whitman (1842-1910). On peut l'appeler " école " puisqu'elle dispose d'un fondateur, d'un lieu où s'exerce sa direction, ici la station de biologie marine fille de l'illustre station de Naples, et d'un groupe de collègues qui utilisent les mêmes techniques. Cette école sera un centre régulier de communications, de discussions, d'exposés théoriques et d'un nombre considérable de travaux expérimentaux. Elle dispose d'une publication les *Biological Lectures,* et d'un périodique mensuel de très haute tenue *The journal of morphology,* fondé par Ch.O. Whitman en 1888. Parmi les participants nous citerons seulement ici Edmund B. Wilson, Frank K. Lillie (1870-1947), Edwin G. Conklin (1863-1952), Thomas H. Morgan (1866-1945).

Charles O. Whitman avait mené ses premiers travaux dans le laboratoire du Pr. Leuckart à Munich. Les sangsues étaient l'organisme choisi, en particulier *Clepsine marginata.* La conclusion à laquelle était parvenu Whitman en 1878 est reprise sous la forme d'un programme de recherche en 1887 : *An accurate acquaintance with the principal events of cleavage is indispensable to a clear understanding of the derivation of the germ-layers ; for these layers, and even the more important organs arising from them, can be traced directly to special blastomeres*[3]. Ce qui s'ouvrait ainsi était révolutionnaire ; c'était une méthode nouvelle, une nouvelle façon de penser.

En effet, comme l'a souligné E.B. Wilson en 1892 dans son étude sur le lignage cellulaire de *Nereis* : " la théorie des feuillets germinatifs constitue en fait la fondation sur laquelle la science entière de l'embryologie comparée est fondée ". (*The germ-layer theory forms, in fact, the foundation on which the entire science of comparative embryology is built*)[4].

Qu'un système entier d'organes puisse être suivi à partir d'un seul blastomère paraissait incroyable. C'est ce que relève Wilson : *That an entire system of organs, such as the ventral-cord, or the trunk-nephridia, could be traced back to a single blastomere was a fact so extraordinary that many morphologists, Balfour among them, at first refused to credit Whitman's statements, notwithstanding the fact that the origin of the entire mesoblast from a single cell had been established in a number of cases*[5].

C'est donc en 1892 qu'apparaît le terme *cell-lineage* dans un titre, ainsi que celui de *Cytogeny : It appears to me that the only course open to embryological investigation is to examine more precisely the origin of the gastrula itself ;*

---

3. C.O. Whitman, " Embryology of *Clepsine* ", *Quart. J. Mic. Sci.,* XVIII (1878) ; *J. of Morph.,* 1 (1887), 108.

4. E.B. Wilson, " The Cell-lineage of Nereis. A contribution to cytogeny of the Annelid Body ", *J. of Morph.,* VI (1892), 363. Pour la théorie des feuillets germinatifs, fondement de l'embryologie comparée, cf. *Le Traité d'Embryologie d'O. Hertwig.* Trad. fr. de Ch. Julin, Paris, 1900. Préface.

5. E.B. Wilson, *op. cit.,* 362. F. Balfour (1851-1882) auteur de *Comparative Embryology,* 1880-1881.

to take as a starting-point not the two-layered gastrula, but the ovum. The 'gastrula' cannot be taken as a starting-point for the investigation of comparative organogeny unless we are certain that the two layers are everywhere homologous. Simply to assume this homology is simply to beg the question. The relationship of the inner and outer layers in the various forms of gastrulas must be investigated not only by determining their relationship to the adult body, but also by tracing out the cell-lineage or cytogeny of the individual blastomeres from the beginning of development...[6].

Ainsi se définissait ce qu'en 1898, dans une étude célèbre *Cell lineage and ancestral reminiscence*, Wilson a appelé : *The cell-lineage program : A long series of later researches, beginning with Whitman's epoch making studies on the cleavage of Clepsine has demonstrated analogous facts in the case of many other cells of the cleaving ovum, and has finally shown that in many groups of animals (though apparently not in all) the origin of the adult organs may be determined cell by cell in the cleavage stages ; that the cell lineage thus determined is not the vague and variable process it was once supposed to be, but is in many cases as definitely ordered a process as any other series of events in the ontogeny ; and that it may accurately be compared with the cell-lineage of other groups with a view to the determination of relationships*[7]. Wilson ajoutait : *The study of cell-lineage has thus given us what is practically a new method of embryological research.*

Cette nouvelle façon de considérer et de faire de l'embryologie allait insister sur le lien entre le clivage et la différenciation, renouveler l'étude de l'homologie des structures embryonnaires, dénombrer les différentes formes de clivage, indiquer le destin des cellules. En relation avec ce destin, Edwin Grant Conklin proposait en 1897 et en 1905, les concepts de clivage déterminé et de clivage indéterminé[8].

Dans les années qui suivent le premier quart du XXᵉ siècle l'étude du lignage cellulaire a paru perdre le rôle indiqué par Wilson. L'école de Hans Spemann (1869-1941) parait sur le devant de la scène. Dans son livre *Embryonic Development and Induction* (1938), Spemann, évoquant l'école américaine, regrette de ne pouvoir avec les amphibiens " suivre un groupe distinct de cellules au cours de leur déplacement durant la gastrulation "[9].

6. E.B. Wilson, *op. cit.*, 367.

7. B.H. Willier and J.M. Oppenheimer, " *Biological Lectures*. Woods Hole Mass. 1898 " in Prentice-Hall (ed.), *Foundations of Experimental Embryology*, 1968, 57.

8. E.G. Conklin, " The embryology of *Crepidula*. A contribution to the Cell lineage and early Development of some marine gasteropods ", *J. of Morph.*, XIII (1897), 173 ; " Organization and Cell-lineage of the ascidian Egg. ", *J. Acad. Nat. Sci. of Philadelphia*, t. XIII (1905), 5-119.

9. Ce programme sera en partie réalisé par Walter Vogt en 1929. Une histoire du marquage cellulaire serait illustrée par le travail de N. Le Douarin où des cellules de caille pouvaient être utilisées comme marqueurs dans les recherches sur les interactions tissulaires ou sur les migrations cellulaires au cours de l'ontogenèse. Le marquage est définitif. N. Le Douarin, " Particularités du noyau interphasique chez la caille japonaise ", *Bull. Biol. de France et de Belgique*, 103 (1969), 435-452.

" La découverte des phénomènes de régulation et d'induction chez les échinodermes et les chordés ont concentré l'attention des embryologistes sur les *interactions cellulaires* plutôt que sur le lignage cellulaire comme facteur causal dans la différenciation ", écrit Gunther Stent en 1985[10].

Pourtant le lignage comme méthode et comme enseignement en embryologie n'était pas oublié. Il était apparemment inapplicable au développement des insectes. L'embryologie de la Drosophile était d'ailleurs mal connue en 1920, ...*its complete development and general biology have not been fully described* écrivait encore M. Demerec en 1950[11]. C'est pourquoi l'on doit être attentif aux travaux d'Alfred F. Huettner qui publie, en 1923, une étude sur l'origine des cellules germinales chez la Drosophile. Il note que son travail porte sur le lignage cellulaire des gonades chez les gynandromorphes de la Drosophile[12].

LES GYNANDROMORPHES COMME PROBLÈME ET COMME OUTIL

La distinction entre mâles et femelles, distinction tranchée, est parfois mise en question lorsqu'apparaissent des individus dont certaines parties sont mâles et d'autres parties femelles. Mais l'importance de ces combinaisons *lies in the opportunity they furnish for analysis of the changes in the hereditary mechanisms of sex determination that makes such combinations possible*[13].

Le problème de la détermination du sexe est un problème ancien que nous ne pouvons aborder ici ; de 1914 à 1919, Morgan et ses collaborateurs confirmaient que les gynandromorphes sont dus à l'élimination d'un des chromosomes X lors de divisions cellulaires précoces. Ce que nous retiendrons c'est la conclusion de Morgan et Bridges (1919) : *A striking fact in regard to these gynandromorphs is that the male and female parts and their sex-linked characters are strictly self-determining each developing according to its own constitution. No matter how large or how small a region may be, it is not interfered with by the aspirations of its neighbors, nor is it overruled by the action of the gonad*[14].

Ceci est remarquable. Cette autodétermination va être le fondement de l'idée géniale de Sturtevant. Chaque partie développe des caractères sexuels et

10. G. Stent, " The role of Cell-lineage in development ", *Phil. Trans. R. Soc. London*, B 312 (1985), 4.

11. M. Demerec, *Biology of Drosophila,* New-York, 1950, Préface. *Cf.* l'étude de B.P. Sonneblick, 88.

12. A.F. Huettner, " The origin of the germ cells in Drosophila melanogaster ", *J. of Morph.,* 37 (1923), 385-419.

13. T.H. Morgan and C.B. Bridges, " The origin of Gynandromorphs ", *Carn. Inst. Wash. Publ.*, 278 (1919), 3.

14. T.H. Morgan and C.B. Bridges, *op. cit.,* 112 ; pour l'histoire de la détermination du sexe, *cf.* S.F. Gilbert, " The Embryological Origins of the Gene Theory ", *J. of The History of Biology,* 11 (1978), 307-351.

des caractères mutants selon sa constitution génétique propre, indépendamment du reste de l'individu. Et c'est seulement parce qu'il en est ainsi, soulignera Sturtevant, qu'il a été possible d'utiliser les gynandromorphes dans une étude sur les problèmes du développement[15].

Alfred H. Sturtevant (1891-1970) fut aussi intéressé par le rôle des gènes dans le développement que par leur transmission. En 1923, il publie une étude sur l'hérédité de la direction de l'enroulement chez les *Limnaea* où il met en évidence l'hérédité maternelle[16].

En 1929, c'est la grande étude qui comprend dans le titre : ...*a study of chromosome elimination and of Cell-Lineage*. Il obtint un nombre considérable de gynandromorphes. Après des croisements, les caractères liés au sexe sont identifiables sur les soies. Sturtevant reporte leur distribution sur des miméographes et des dessins qu'Antonio Garcia-Bellido retrouvera quarante ans plus tard. La construction de la carte des destins cellulaires est d'une apparente simplicité, tout comme la carte des gènes en 1913. Comme il était impossible de suivre les clivages nucléaires précoces chez la Drosophile, son raisonnement fut le suivant : les gynandromorphes proviennent de clivages précoces qui comprennent des cellules mâles et des cellules femelles. Celles-là même qui constituent les parties relevées sur les diagrammes.

It is evident that a systematic study of these diagrams can give considerable information as to the cell-lineage of Drosophila, for parts that are genetically closely related will more often be alike as to sex than will parts more remotely related[17].

VERS LA GÉNÉTIQUE DU DÉVELOPPEMENT

Désormais le généticien dispose d'un outil d'analyse dans ses études sur le développement, celui des mosaïques. Michael Ashburner et Walter Gehring se sont étonnés que l'étude de Sturtevant soit restée dans l'ombre pendant quarante ans. Ce qui va changer au début des années 1970, c'est une étude beaucoup plus systématique des clones cellulaires des disques imaginaux considérés comme modèles du développement ainsi que l'analyse génétique des transformations homéotiques accomplies par E.B. Lewis autour des années 1960. On voit apparaître de nouveaux outils dont l'utilité pour l'étude génétique du développement est évidente mais qui ne seront exploités que plus tard.

---

15. A.H. Sturtevant, " The Claret mutant type of *Drosophila simulans*. A study of chromosome elimination and of Cell-lineage ", *Zeit. f. Wissen. Zool.*, 135 (1929), 323-356, et *Selected Papers of A.H. Sturtevant, Genetics and Evolution*, Ed. Lewis, ed. W.H. Freeman, 1961, 142.

16. A.H. Sturtevant, " Inheritance of direction of coiling in Limnaea ", *Sci.*, 58 (1923), 269-270.

17. A.H. Sturtevant, *op. cit.*, 1929, 132.

Ainsi, J.T. Patterson obtient des *mutations somatiques* à l'aide de rayons X[18].
En 1936, après de longues études génétiques, Curt Stern démontre l'existence
d'un *crossing over somatique*[19]. Découverte qui va jouer un rôle essentiel dans
l'étude du lignage puisque chaque cellule sera marquée et suivie.

Nous ne pouvons nous arrêter sur les études nombreuses qui, autour de
1940, vont orienter les travaux vers une approche génétique du développement.
Nous ne pouvons que rappeler la publication de *Physiological Genetics* (1938)
de R.B. Goldschmidt ; la première étude de Salomé Glucksohn-Schoenheimer
sur le locus T de la souris ; Charlotte Auerbach étudie les mutations de l'aile
de la Drosophile. L'aile, proposée comme modèle du développement en 1939,
par C.H. Waddington. Enfin, D.F. Poulson étudie en 1940 l'effet de l'absence
d'un gène, donné sur le développement embryonnaire de la Drosophile[20].

La période qui va de 1950 à 1975 sera illustrée, pour notre propos, par les réflexions et les travaux de Curt Stern, Ernst Hadorn et Antonio Garcia-Bellido

L'apport décisif de C. Stern a été d'appliquer une notion essentielle pour
l'analyse génétique du développement : celle de l'*autonomie* de l'expression
génétique de chaque cellule. Cependant, l'autonomie pour lui est une *réponse*,
réponse à un *pattern* invisible, prédéterminé, qui est l'élément décisif. Ainsi se
présente le concept de *prepattern,* équivalent du concept de champ embryonnaire des embryologistes[21].

Les travaux de l'école de Hadorn à l'université de Zürich demandent une
étude à part. Les études de transplantation se sont déroulées à partir de 1946.
A partir de 1962, Hadorn et ses collaborateurs avaient réussi à conserver des
populations de cellules dans des cultures permanentes. Quant au phénomène de
transdétermination, il montrait que les cellules de disques imaginaux peuvent
changer leur détermination et se différencier selon des structures qui correspondent à d'autres disques.

18. J.T. Patterson, " Somatic Mutations induced by X-rays ", *J. of Heredity,* 20 (1929), 261-
267.
19. C. Stern, " Somatic crossing-over and segregation " in *Drosophila melanogaster. Genetics,*
21 (1936), 626-728.
20. Pour les travaux de Salomé Glucksohn-Schoenheimer, *cf.* S.F. Gilbert, *A conceptual History of Modern Embryology,* The Johns Hopkins Univ. Press, 1994, chap. 9. Charlotte Auerbach,
" Wild type and some mutant strains " in *Dros. melan. Trans. R. Soc. Edin.,* 58 (1935-1936), 787-
815 ; C.H. Waddington, " The Genetic control of wing development ", *J. of Genetics,* 41 (1940),
75-137 ; D.F. Poulson, " The Effects of certain X-chromosome deficiencies of the embryonic
development of Dros. melan. ", *J. Exp. Zool.,* 83 (1940), 271-325.
21. C. Stern, " Two or three Bristles ", *The American Scientist,* 42 (1954), 242, and *Genetic
Mosaics and other Essays,* Harvard Univ. Press., 1968, chap. 3.

Le dilemme qui se pose en 1966 peut être résumé ainsi : les cellules sont-elles subordonnées à un champ et se différencient-elles selon le système supracellulaire auquel elles appartiennent ? Ou bien les cellules sont-elles spécifiquement déterminées ?[22] Les expériences de dissociation et de réagrégation de cellules génétiquement marquées faisaient pencher vers la seconde hypothèse. Les expériences de W. Gehring pouvaient être interprétées ainsi : la transdétermination peut se produire simultanément dans des groupes de plusieurs cellules. Celles de Garcia-Bellido montraient que l'affinité ou l'absence d'affinité des cellules plaident en faveur d'une différenciation cellulaire autonome.

Ainsi apparaît la question centrale : comment l'information génétique des cellules individuelles est-elle exprimée dans une dimension supra-cellulaire ?

Or, le terrain expérimental qui peut apporter une réponse à cette question est celui des clones cellulaires — chez la Drosophile, il s'agit en fait de *polyclones* — c'est-à-dire du lignage. D'où la première communication d'A. Garcia-Bellido dès son arrivée chez Ed. Lewis au California Institute of Technology : " Le lignage cellulaire dans le disque de l'aile de la Drosophile "[23]. Cette même préoccupation le conduit avec John Merriam à reprendre l'admirable étude de Sturtevant (1929)[24].

Nous ne pouvons qu'indiquer les directions qui concourent à la conclusion.

L'étude comparée des travaux de Schneiderman et de ses collaborateurs à Irvine en Californie[25] avec ceux de l'école de Madrid, montre que les premiers vont mettre en évidence la *séparation* réelle des lignages cellulaires dans le développement de l'aile ou de la patte chez la Drosophile. Les seconds vont être attentifs à la ligne de partage des lignages. Il s'agira désormais de lignages cellulaires *restreints*. Les bords de ce que Garcia-Bellido appellera compartiment, unité supra cellulaire du développement (1973) seront mis en évidence grâce aux mutations *engrailed* et *Minute*.

…We postulate the existence of a splitting mechanism segregating different groups of cells from previously homogeneous and contiguous cell population. These groups will show restricted cell lineages, defining developmental compartments in their subsequent development, visualised in the adult cuticle by fixed clonal demarcation lines[26].

22. E. Hadorn, " Dynamics of Determination " in M. Locke (ed.), *Major Problems in Developmental Biology,* Acad. Press, 1966, 94-95.

23. A. Garcia-Bellido, " Cell-lineage in the wing disc of Dros. melan. ", *Genetics,* 60 (1968).

24. A. Garcia-Bellido and J.R. Merriam, " Cell-lineage of the imaginal Discs in Drosophila Gynandromorphs ", *J. Exp. Zool.,* 170 (1969), 61-75.

25. P.J. Bryant and H.A. Schneiderman, " Cell-lineage growth and determination on the imaginal discs of *Dros. melan.* ", *Dev. Biol.,* 20 (1969), 263-290.

26. A. Garcia-Bellido, P. Ripoll, G. Morata, " Developmental compartimentalisation of the wing disc of Drosophila ", *Nature New Biol.,* 245 (1973), 251-253.

En bref, on pouvait désormais fondre en une seule démarche l'analyse génétique des mutations et l'analyse clonale[27]. Un des derniers paradoxes est que
cela ait été obtenu avec la Drosophile en prenant pour modèle le disque imaginal. L'étude du lignage chez le nématode ne sera reprise qu'immédiatement
après.

NOTE

Depuis cette communication deux importantes publications doivent être
signalées :

Alain Ghysen (ed.), " Developmental Genetics of *Drosophila* ", *The International Journal of Developmental Biology*, Special issue, vol. 3 (1998).

A. Garcia-Bellido, " The Engrailed Story ", *Genetics*, 148 (1998), 539-544.

27. A. Garcia-Bellido and P. Ripoll, " Cell-lineage and Differenciation " in W.J. Gehring (ed.),
Genetic Mosaics and Cell Differentiation, Springer, 1978.

FROM EXPERIMENTAL EMBRYOLOGY TO DEVELOPMENTAL BIOLOGY : THE METAMORPHOSES OF THE GRADIENT CONCEPT

Denis THIEFFRY

Despite recurrent claims and promises of several pioneering molecular biologists, the problem of embryonic development proved to be particularly intractable within the molecular framework. Accordingly, many tensions arose between the new discipline and more traditional embryological fields throughout the 1960s and 1970s. During the last decade, critics largely faded out, leading to an almost complete hegemony of molecular biology over embryological and developmental problems. This hegemony is often considered as the result of a long-awaited but final resolution of the main molecular mechanisms at the basis of developmental processes. Also, it involves the molecular translation or re-interpretation of many embryological concepts, such as " induction ", " regulation ", " homeosis ", or " epigenesis ". However, perhaps surprisingly, some of the main breakthroughs that led to our present molecular understanding relied on rather classical approaches, i.e. methods, systems and techniques borrowed from classical genetics or experimental embryology. In this paper, I focus on the history of an important embryological concept that was only recently integrated into the molecular conception of development, the concept of " gradient ". Such historical reconstruction and especially the tracing of semantic shifts serve here as a lens into how the study of embryogenesis was redefined so that it could be integrated into the new molecular framework.

The notion of gradient can be traced back at least to early developments of experimental embryology at the beginning of this century[1]. Indeed, a " graded distribution of material " was already invoked by Morgan in 1905, to account for the polarity effect observed in regeneration experiments on a variety of

---

1. Several essays already addressed various aspects of the history of embryology and regeneration research ; see, e.g., T.J. Horder, J.A. Wikowski and C.C. Wylie (eds), *A History of Embryology*, Cambridge, Cambridge University Press, 1986 ; S.F. Gilbert (ed.), *A Conceptual History of Modern Embryology*, New York, Plenum Press, 1991 ; C.E. Dinsmore (ed.), *A History of Regeneration Research*, Cambridge, Cambridge University Press, 1991.

organisms, including *Tubularia* and *Lombricus*[2]. A similar notion, coined *Gefälle* in 1910, was developed by T. Boveri in the context of his conception of a stratification of the egg along the animal-vegetative axis[3]. Whereas neither Morgan nor Boveri made precise assumptions about the physical nature of these gradients, C.M. Child conceived this gradient as one in metabolism[4]. Publicised in various papers and books, this conception was picked up by many embryologists between 1910 and 1930. Among these were J.S. Huxley and G.R. de Beer, who coined the term " gradient field ", and the Belgian Albert Dalcq, who, in his *L'Oeuf et son Dynamisme Organisateur*, emphasised the role of thresholds (*seuils*) in relation with gradients and fields (*champs*)[5]. Gradients enjoyed significant popularity among embryologists at least up to the 1940s. However, some embryologists of that time became quite critical about the gradient hypothesis, especially in its special metabolic form. One of the most influential critics was Hans Spemann, who devoted a whole chapter of his *Embryonic Development and Induction* to a critical analysis of the gradient theory[6].

From the mid 1940s, gradients progressively lost their seductive power. The oblivion became so great that barely any entry for *gradient* can be found in the indexes of the embryological treatises of the 1970s. However, some specific groups continued to refer to gradients, especially people working on regeneration and on *Drosophila* development, as well as some theoretical biologists[7].

At the end of the 1960s, Lewis Wolpert, an engineer turned biologist and working on hydra regeneration, proposed an abstract representation of the setting of spatial organisation. With his " French Flag Problem ", Wolpert meant to model pattern formation, with special reference to both size invariance and regeneration properties. In this context Wolpert introduced the concept of " positional information ", covering various types of mechanisms, including gradients and reaction-diffusion mechanisms *à la* Turing[8]. First presented at the 1966 Theoretical Biology Conference organised by Waddington in Bellagio, Wolpert's idea was sympathetically received. Although the audience

2. T.H. Morgan, *J. Exp. Zool.*, 2 (1905), 495-506.

3. T. Boveri, *Arch. Entw. Mech.*, 30 (1910), 101-125.

4. C.M. Child, *Science,* 39 (1914), 73-76.

5. G.R. De Beer and J.S. Huxley, *The Elements of Experimental Embryology*, Cambridge, Cambridge University Press, 1934 ; A. Dalcq, *L'Oeuf et son Dynamisme Organisateur*, Paris, Albin Michel, 1941.

6. H. Spemann, *Embryonic Development and Induction*, New Haven, Yale University Press, 1938.

7. A notable exception is the work of S. Toivonen and collaborators, who kept the notion of gradients in the embryological literature through the nineteen sixties. For a recent account, see S.F. Gilbert and L. Saxen, *Mech. Dev.*, 41 (1993), 73-89.

8. A.M. Turing, *Phil. Trans. Roy. Soc. London*, Ser B, 641 (1952), 37-72. In this paper, Turing presented an extensive mathematical treatment of 2- and 3- dimensional reaction-diffusion models accounting for the formation of various spatio-temporal patterns from homogenous initial conditions. However, simple gradients were not specifically addressed by Turing.

proved to be more critical at Woods Hole later in the same year, this did not prevent Wolpert to publish an extensive review of his ideas on the relation between "Positional information and the Spatial Pattern of Cellular Differentiation " in the *Journal of Theoretical Biology* in 1969[9].

As a consequence, Wolpert triggered the interest of one of the most influential molecular biologists of the time, namely Francis Crick. Indeed, Crick published a short paper on " Diffusion in Embryogenesis " in *Nature* the following year. In this paper, Crick took seriously the idea of a molecular gradient and diffusion as a sound mechanism for the establishment of gradients. Writing a few simple equations to account for a diffusion-based formation of a gradient, he crudely estimated some of the main parameters, e.g., time scale, viscosity, molecule size, and spatial range, reaching figures that were in reasonable agreement with the data reported in Wolpert's review. Despite evident over-simplifications, Crick's paper contributed " to make the idea of diffusion gradients respectable [again] to biologists "[10].

At this stage, gradients regained the interest of several research groups, especially theoretical biologists. The gradient concept was reinterpreted in order to fit a quantitative representation in terms of (Cartesian or polar) spatial co-ordinates. However, gradients were then usually contemplated as one mechanism among others accounting fot pattern formation, and were thus often considered as a subordinate means of providing positional information.

Further theoretical developments principally addressed the problems of gradient establishment and interpretation. It will suffice to mention here the 1977 contribution of Lewis *et al.,* who invoked " positive autoregulation " as a mechanism accounting for both discontinuous interpretation and memorisation of positional value, and Hans Meinhardt's sophisticated models for the establishment of orthogonal axes and compartment boundaries[11].

At the same time, a few experimental groups remained interested in the gradient hypothesis. Wolpert himself, as well as Gierer (collaborating with Meinhardt) frequently invoked gradients to account for the results of their regeneration experiments. Other works, involving less theoretically inclined biologists, dealt with the experimental analysis of insect development. Strikingly, these works relied mostly on classical genetic and embryological techniques at a time of overwhelming spread of molecular techniques. For example, Klaus Sander was led to propose a model involving two opposite antero-posterior gradients from the results of a series of experiments consisting

9. L. Wolpert, *J. Theor. Biol.,* 25 (1969), 1-47. Wolpert published several personal accounts on the history of gradient, positional information, and pattern formation ; see. e.g., his essays in Horder *et al.* (eds) and in Dinsmore (ed.), mentioned in footnote 1.

10. F.H. Crick, *Nature,* 225 (1970), 420-422.

11. J. Lewis, J.M. Slack and L. Wolpert, *J. Theor. Biol.,* 65 (1977), 579-590 ; H. Meinhardt, *Models of Biological Pattern Formation*, New York, Academic Press, 1982.

of ligature of eggs and transplantation of cytoplasm in a leaf-hopper, following the roads opened by the work of earlier experimental embryologists[12].

Similarly, H.F. Stumpf used standard embryological techniques (transplantation and rotation of pieces of skin) to support the thesis following which a concentration gradient is involved in the positioning of cuticular structures in various insects (*Galleria* and *Rodnius*)[13].

Still more striking is the fact that the most important breakthrough was provided by a molecularly trained biologist, C. Nüsslein-Volhard, who used rather traditional techniques, e.g., mutagenesis, genetic crossing, and embryo staining techniques. In the mid 1970s, Nüsslein-Volhard and E. Wieschaus started a systematic search of embryonic lethal mutants. Encompassing a whole decade, this gigantic work led them to find dozens of genes involved in the formation of *Drosophila* early embryonic pattern. These mutants were distributed into four classes: "maternal", "gap", "pair-rule" and "segment-polarity", according to their phenotypes[14]. In particular, a series of maternal mutants were involved in the laying down of antero-posterior and dorso-ventral gradients, providing the basic heterogeneity on which other genes could progressively build the characteristic stripe pattern. At this stage, molecular biologists entered into play, identifying gradients with specific macromolecules (i.e., the products of bicoid, hunchback and nanos genes for the antero-posterior axis, nuclear dorsal protein for the dorso-ventral axis), decrypting the mechanisms by which the gradients are established, and isolating and sequencing the corresponding genes and their regulatory regions.

Today, hundreds of papers, published in renowned journals, deal with molecular gradients and their roles in the formation of various patterns across several different phyla. Certainly, molecular biologists recognise that there is still a lot of work to do in order to reach a satisfactory account of the establishment and interpretation of gradients. However, they generally agree on the fact that gene regulatory networks must somehow be involved at the core of this picture. In addition, as is particularly clear from the title of a recent review stating that " Fish are like flies are like frogs "[15], molecular biologists are very keen to emphasise the deep unity of basic patterning mechanisms involved in early embryonic development. Thanks to their molecular identification, gradients have gained full citizenship in the new developmental biology.

Summarising, the gradient concept has followed a long and torturous path since its original expression in the context of experimental embryology, pass-

12. K. Sander, *Ciba Found. Symp.*, 29 (1975), 241-263. See also Sander's historical essay in Horder *et al.*, footnote 1.

13. See e.g. H.F. Stumpf, *Nature*, 212 (1966), 430-431.

14. C. Nüsslein-Volhardt and E. Wieschaus, *Nature*, 287 (1980), 795-801 ; C. Nüsslein-Volhardt and S. Roth, *Ciba Found. Symp.*, 144 (1989), 37-55. For a circonstancied historical analysis of Nüsslein-Volhardt's work, see E.F. Keller, *Hist. Stud. Phys. Biol. Sci.*, 26 (1996), 313-346.

15. S.A. Holley and E.L. Ferguson, *BioEssays*, 19 (1997), 281-284.

ing through various re-interpretations, by experimental embryologists, theoretical biologists, developmental geneticists, and finally by contemporaneous molecular embryologists. In the meantime, emphasis shifted from global properties of the developing egg or of the regenerating organ, to specific genes, RNA's, or proteins, before ultimately shifting back towards a more global perspective in terms of gene networks.

On the basis of this preliminary overview, I would suggest that more extensive studies of long-standing concepts like that of the gradient could contribute to a better understanding of the various semantic shifts concomitant with the development and extension of the new molecular paradigm.

I would like to thank Charles Galperin and Scott Gilbert for their kind invitation to participate in this stimulating session devoted to the *New Biology of Development*, as well as Lily Kay, Richard Burian, Hans-Jörg Rheinberger, and Evelyn Fox Keller for several helpful comments and suggestions.

ALBERT DALCQ (1893-1973), EMBRYOLOGISTE THÉORICIEN

Jean-Louis FISCHER

A. Dalcq appartient à la grande école d'embryologie belge qui tient ses origines avec Edouard Van Beneden (1826-1910), découvreur de la méiose et du centrosome[1]. L'un des maîtres de Dalcq en embryologie fut Albert Brachet (1869-1930), lui-même formé à la discipline par Van Beneden. A. Brachet, fondateur du laboratoire d'embryologie humaine à l'université de Bruxelles, considérait que l'objectif de la science de l'oeuf et de l'embryon est " … L'étude des problèmes fondamentaux du développement " qu'il décrit dans *La vie créatrice des Formes* (1927). C'est l'objectif qui sera suivi par Dalcq, successeur de Brachet à la direction du laboratoire d'embryologie de Bruxelles en 1935. Deux des principaux collaborateurs de Dalcq qui chercheront avec lui à découvrir et expliquer le fonctionnement du germe dans son dynamisme et son chimisme sont Jean Pasteels et Jean Brachet.

Dans l'oeuvre embryologique de Dalcq, riche en faits, en découvertes et dans ses positions théoriques, deux aspects nous semblent mériter une attention particulière dans la mesure où ils expriment sa pensée dans ses conceptions expérimentales et didactiques. Il s'agit de son " Étude des localisations germinales dans l'oeuf vierge d'ascidie par des expériences de mérogonie " (1932)[2] et de *L'oeuf et son dynamisme organisateur* (1941)[3].

L'étude sur les localisations germinales, effectuée par Dalcq, n'est que l'aboutissement d'une suite de travaux et de réflexions qui ont été à l'origine même de l'embryologie expérimentale. Les premiers travaux fondateurs sont ceux de Chabry (1887)[4]. Chabry, après avoir inventé et adapté une technique

---

1. G. Hamoir, *La découverte de la méiose et du centrosome par Edouard Van Beneden*, Académie royale de Belgique, 1994, 128 p.

2. A. Dalcq, " Étude des localisations germinales dans l'oeuf d'ascidie par des expériences de mérogonie ", *Arch. d'Anat. Microsc.*, t. XXVIII, fasc. II (Juillet 1932), 224-333.

3. A. Dalcq, *L'oeuf et son dynamisme organisateur*, Paris, Albin Michel, 1941, 582 p.

4. L. Chabry, *Contribution à l'embryologie normale et tératologique des ascidies simples*, Paris, Félix Alcan, 1887, 154 p.

pour réaliser son projet expérimental[5], démontrait que l'oeuf d'ascidie présente des structures préformées aux stades de deux et quatre blastomères. Toutefois si les expériences de Chabry ouvraient d'une façon définitive le champ de l'expérimentation en embryologie, elles conduisaient également à repenser et à réactiver le débat entre préformation et épigenèse. La preuve en était apportée par les interprétations divergentes qui étaient faites au sujet des résultats présentés par Chabry. W. Roux appliquait à l'oeuf d'ascidie sa conception de la " post génération " (1892) qui était le compromis permettant une conciliation entre préformation et épigenèse ; D. Barfurth, dans le cadre de la néo-préformation, considérait l'oeuf d'ascidie comme étant bien anisotrope (mosaïque) ; et O. Hertwig et H. Driesch (1894), défendant le concept de la néo-épigenèse, déduisaient des expériences de Chabry que l'oeuf d'ascidie était isotrope (régulation)[6].

En 1917 A. Brachet, rappelant les premières expériences de 1905 de R.G. Conklin sur le développement mosaïque de l'oeuf d'ascidie et son anisotropie entre le moment de la fécondation et son premier clivage, notait que si les expérimentateurs n'avaient jamais pu faire produire à l'oeuf d'ascidie " plus que ce qu'il fait " pour la raison que " sa potentialité réelle semble… se confonde avec sa potentialité totale ", il ne fallait pas pour autant en conclure à la rigidité absolue de ses localisations germinales. Le problème avait déjà fait l'objet de réponses à propos de l'oeuf de grenouille considéré comme anisotrope mais qui pouvait réguler dans certaines conditions expérimentales (T.H. Morgan. 1895).

C'est entre le 15 juillet et le 15 septembre 1931 au laboratoire de la station maritime de Roscoff, dirigé par C. Pérez (directeur) et G. Teissier (sous-directeur), que Dalcq va effectuer une série d'expériences sur l'oeuf d'une ascidie *(Ascidielle aspersa),* qui avait servi de matériel expérimental à Chabry. L'idée poursuivie par Dalcq dans ces expériences est de révéler si l'oeuf d'ascidie, considéré comme parfaitement anisotrope, possède des facultés régulatrices avant la fécondation. S'inspirant d'une récente publication de G. Reverberi sur la fécondation d'extraovats obtenus expérimentalement sur des oeufs d'une ascidie, *Ciona intestinalis,* Dalcq pense utiliser dans sa méthode expérimentale celle de la mérogonie inventée par Y. Delage en 1899. La mérogonie consistant à féconder les morceaux cytoplasmiques nucléés et non nucléés découpés dans un oeuf vierge. La fécondation par Reverberi d'un extraovat isolé correspond

5. J.L. Fischer, consulter les études sur Chabry in K. Sander in coll., *Landmarks In developmental Biology 1883-1924*, Berlin ; Heidelberg, Springer-Verlag, 1997.

6. A ce propos A. Brachet écrit dans son ouvrage *L'oeuf et les facteurs de l'ontogenèse*, Paris, Doin, 1917 : " Il conviendrait peut-être chez les Ascidies de ne porter son attention, quand on étudie la potentialité des blastomères, que sur les stades gastrula, neurula et chordula. En effet, quand la larve urodèle se forme, il devient très difficile de dire, en raison de sa structure même, si elle est vraiment partielle, et surtout jusqu'à quel point elle l'est. C'est d'ailleurs la raison pour laquelle les observations de Chabry ont pu être interprétées de diverses façons. ", p. 273, note 1.

à une mérogonie : seule la méthode expérimentale diffère de celle de Delage, elle-même étant différente de celle de Th. Boveri (1889) qui pratiquait la méthode du secouage de l'oeuf vierge dans un tube pour le faire éclater en plusieurs fragments.

Dalcq réalise de nombreuses expériences de fécondation d'oeufs vierges mérogones d'ascidie, sectionnés en deux ou, plus rarement, en trois segments, suivant des axes méridiens ou équatoriaux, et dont les parties nucléées et non nucléées sont plus au moins égales. Il est attentif à tous les résultats, fidèle à sa conviction d'embryologiste le conduisant à défendre cette position que " Dans l'ensemble du problème de la morphogenèse, la notion des localisations germinales est la base de toute explication. Quelle que soit la part qui revienne aux processus d'induction dans l'organisation de l'embryon, ceux-ci restent secondaires vis-à-vis des différences primordiales entre les diverses régions du cytoplasme de l'oeuf "[7].

Les résultats sont nombreux et apportent de nouvelles connaissances concernant la régulation de l'oeuf d'ascidie. Sur 61 expériences réussies ayant fait l'objet d'une analyse approfondie, Dalcq apporte ces conclusions nouvelles : l'oeuf vierge d'ascidie possède des localisations germinales réparties suivant un plan de symétrie bilatérale ; lorsque la scission de l'ovule passe très précisément par le plan méridien on obtient deux larves jumelles " harmonieusement constituées ", et certains fragments d'oeuf produisent une larve normale. Ces derniers résultats laissent à penser que, dans des conditions expérimentales précises, l'oeuf d'ascidie a un pouvoir de régulation. Toutefois, sans nier un évident phénomène régulatoire de l'oeuf d'ascidie, résultat paradoxal pour un oeuf reconnu comme l'exemple type de l'oeuf anisotrope, et induit par le geste de l'expérimentateur, Dalcq avertit qu'il ne faut pas pour autant considérer l'oeuf d'ascidie comme appartenant aux oeufs à régulation. Au contraire, précise-t-il ses " observations montrent … que les localisations sont parfaitement définies dans l'oeuf vierge mais avec une disposition qui leur confère néanmoins quelque souplesse "[8]. Cette " souplesse ", chez un oeuf mosaïque ne pouvant produire que les potentialités que son phylum lui a léguées, dévoilée et reconnue par Dalcq, va le conforter dans sa prise de position que " le développement dans son ensemble apparaît comme une vaste épigenèse, mais sa compréhension exige néanmoins, au départ, un minimum de préformation "[9]. Cette façon de penser sera particulièrement critiquée par Paul Vintrebert qui militait pour un développement exclusivement épigénétique ; et

7. A. Dalcq, *loc. cit.*, note (2), 224.

8. *Ibid.*, 332.

9. A. Dalcq, " Préformation et Epigenèse dans leur acception actuelle (avec la notion d'Epigenèse trophique) ", *Bulletin de l'Académie royale de Belgique*, classe des sciences, 1953, t. 39, 1139.

Dalcq ne manqua pas de répondre aux attaques souvent plus idéologiques que scientifiques de l'auteur du " développement du vivant par lui-même "[10].

Dalcq ne s'est pas cantonné dans la seule recherche expérimentale en embryologie. Enseignant, il a joué un rôle important dans la diffusion des connaissances en embryologie. Vulgarisateur de sa discipline, il a publié plusieurs ouvrages de grande synthèse dont *L'oeuf et son dynamisme organisateur* en 1941. Cet ouvrage n'est certes pas destiné à un large public ; il vise plutôt les médecins et biologistes plus ou moins étrangers à l'embryologie et qui pourront " rénover leurs conceptions à la lumière des révélations les plus récentes de l'Embryologie " ; ce livre s'adresse aussi à tous ceux " qui s'intéressent à l'évolution générale de la pensée scientifique ". L'un des buts que se propose donc Dalcq, dans la publication de ce livre, est de rendre accessible l'embryologie aux scientifiques non spécialisés dans la science du développement ; parce que cette science s'est isolée des autres disciplines biologiques en raison d'une terminologie et d'un langage " d'apparence hermétique pour qui n'a pas eu l'occasion d'observer les phénomènes auxquels ces termes s'appliquent ". L'autre but poursuivi par Dalcq est de " montrer l'évolution générale des idées " et de " dégager la dialectique actuelle de l'Embryologie " et non d'écrire un traité technique d'embryologie.

Livre d'informations, de débats théoriques, défendant la " rationalisation des phénomènes " et la " conviction que le principe de causalité est l'agent de liaison indispensable entre notre esprit et le monde extérieur ", *L'oeuf et son dynamisme organisateur* est la confession par son auteur de ces longs moments de " silencieuses conversations avec le germe en développement ". L'organicisme de Dalcq, proche de celui de Bertalauffy et Voodger, veut que " l'organisation intrinsèque du protoplasme " soit " la caractéristique de la vie et le but essentiel de l'investigation biologique " ; mais Dalcq connaît aussi les limites de la vérité de l'instant et, ouvert aux premiers résultats de la génétique du développement, il conçoit que " des perspectives nouvelles s'entrouvrent sans cesse " et que des " découvertes plus formelles fermeront les unes et élargiront les autres ". Oeuvre scientifique et littéraire, cette synthèse des connaissances embryologiques de la fin des années 1930, donne toute la mesure de la réflexion de l'embryologiste, quand l'interprétation expérimentale côtoie l'extrapolation philosophique et la sensibilité poétique engendrées par la passion de la découverte du vivant en mouvement.

---

10. Au sujet de cette querelle consulter A. Dalcq, " La régulation dans le germe et son interprétation ", *C. R. Soc. Biol.*, t. 119, N° 28 (1935), 1421-1466, et la réponse de Vintrebert p. 1466-1480. Puis P. Vintrebert, *Le développement du vivant par lui-même*, Paris, Masson, 1963, 455 p.

BORIS EPHRUSSI ON THE UNITS
OF INHERITANCE AND OF DEVELOPMENT

Richard M. BURIAN

The essential thing we are missing here is not so much a complete catalogue of the ultimate parts that make up different organisms, as the understanding of the way these parts are assembled and integrated into structures of successive, higher levels of organisation. This is the object of developmental biology, where the unit we start with and from which we build up is the cell, the lowest biological unit endowed with the integrating properties of an organism... Here, among the first things that face us, is the famous paradox of cell differentiation, i.e., the problem of the mechanism whereby the descendants of a single cell, all endowed with complete and identical sets of genetic material, acquire widely different and often very stable characteristics. Classical genetics... could provide no explanations for the emergence, *in a temporally and spatially ordered fashion,* of different stable biochemical characteristics, and of their maintenance, on a common genetic background. That is why many geneticists rejected the problem of cell differentiation as not belonging to their province, and why some others (I for one), insisted that it must find its explanation in the inheritance of cytoplasmic variations, many examples of which have indeed been round, especially among micro-organism. [From a lecture delivered in October, 1962 at Western Reserve ; published posthumously as (Ephrussi, 1979), p. 693].

I. Boris Ephrussi (1901-1979) died before molecular developmental biology in its current form had taken shape. Thus, he cannot count as a leading figure of the new developmental biology. Yet he unquestionably deserves to count as one of the forefathers of the new discipline and his career illuminates the transformations that brought that field into being. He pioneered the use of key techniques and brought a fascinating mixture of ingredients, many of which are central to the ongoing reorientation of work in developmental biology, to bear in his work. The conceptual apparatus he developed still provides useful insights into the problems and issues faced by developmental biology. The key point, perhaps, is that although he was, in the end, a geneticist, he began his career as an experimental embryologist with strong cytological interests and always retained the focus on problems of development. The problems that most fascinated him, which he came to attack by use of genetic tools, were thoroughly embryological in character. He came to believe very strongly that organismal development could not be understood except via genetics — and (eventually) traditional nuclear genetics at that. His insistence on using the tools of genetics and on the claim that, nonetheless, genetics alone can not handle embryological problems may seem paradoxical. In the end, however, the result was an enormously fruitful recognition that the solution of such problems as the control of cellular differentiation must depend, in ways we will explore briefly in this paper, on the behavior of cells and of organisms as integrated entities. In sum, the problems he tackled arose from the concerns of a developmental biologist rather than those of a geneticist but, after World War II, he approached *development* by employing the tools of genetics. His work provides many clues regarding key features of the field that eventually became developmental genetics[1].

I shall argue that the key unifying features in Ephrussi's work can help us to understand the new developmental biology. For this reason, it may be useful to remind you, very briefly, of the great diversity of the projects at the center of his career. The list that follows reinforces the point that his main projects reflect his predominantly embryological or developmental orientation. Here are the five major programs of research he undertook, all impressively successful.

1. *Ca.* 1922-1932. Study of sea urchin development from unfertilized eggs up to the 40-hour pluteus larva. His program included analysis of the changing chemical composition of the egg, the larva, and the pluteus, with special concern for the control of development in response to what we now call heat shock.

2. *Ca.* 1928-1935. The development of tissue culture techniques in order to understand the specific properties of differentiated cell types, the conditions under which differentiated properties are expressed, and the integration of tissues (i.e. the extend to which a tissue responds as a whole to a trauma or to a change in its local environment). Here, the integration of tissues refers to the fact that, in various circumstances to which the tissue in a culture appears to respond as a whole (e.g., by resuming its original shape in wound healing), different cells within the culture, demonstrably identically committed or determined, express different properties. The integrated behavior of tissues cultured *in vitro* so impressed Ephrussi that he treated it as a model of the integration

---

1. This paper builds on prior work, in collaboration with Jean Gayon and Doris Zallen. See Burian, *et al.*, 1988, esp. 389-400 ; Burian, 1990 ; Burian and Gayon, 1990 ; Burian and Gayon, 1992 ; Burian, *et al.*, 1991 ; Burian and Zallen, 1992 ; Zallen and Burian, 1992 ; and Gayon, 1994. Other useful work on Ephrussi may be found in Lwoff, 1979 ; Roman, 1980 ; Roman, 1982 ; Roman, 1986 ; Sapp, 1986, chap. 5 ; Sapp, 1986 ; and Fincham, 1992.

of an organism[2]. An important extension of this work was a study of the properties of cells and tissues from embryos of mice carrying the (homozygous lethal) t-allele. Ephrussi demonstrated that fibroblasts taken from the lethal homozygotes could differentiate to form cartilage, a tissue not produced in the affected embryos since they all died 3-4 days before the formation of cartilage. He argued on the basis of his experiments that the defect involved was a failure of integration of the organism rather than a cellular defect or some defect in a particular tissue. (See Ephrussi, 1933, Ephrussi, 1935).

3. *Ca.* 1934-1945. The control of non-autonomous properties in *Drosophila,* such as eye color, and expression of *bar, scute, cinnabar, and vermilion,* studied by transplantation experiments in his famous collaboration with Beadle. As we (Burian *et al.*, 1991 ; Zallen and Burian, 1992) and Scott F. Gilbert (1991) have argued, this work employed *embryological* techniques on genetically defined stocks in order to investigate the role(s) of genes in ontogeny. The work drew on the *Ephestia* work of Kühn and his colleagues discussed by Rheinberger in this symposium. It resulted in careful analysis of the roles of the diffusible substances produced under the control of the *vermilion* and *cinnabar* genes. This project led to the analysis of the metabolic pathways involved in eye pigment production, followed, eventually, by biochemical analysis of the compounds involved, which the German biochemists solved ahead of both Beadle's and Ephrussi's groups. This line of experiments led fairly directly to Beadle and Tatum's *Neurospora* work and the one gene — one enzyme hypothesis. (See the references in n. 1 for guidance to this literature. Ephrussi's own reviews of the work may be found in Ephrussi, 1942, Ephrussi, 1942).

4. *Ca.* 1946-1960. Study of the *petite colonie* mutation in yeast. As is well known, this mutation occurs spontaneously in wild-type *Saccharomyces cerevisiae* at about 0.1 - 0.2% and involves the irreversible loss of the ability to respire. Ephrussi and his colleagues showed that this condition stems from the loss of an entire suite of enzymes (including certain cytochromes) and is due to the fact that certain cytoplasmic particles, eventually firmly shown to be mitochondria, lose the ability to produce those enzymes. Since some acridine dyes could cause 100% loss of respiratory competence without other damage

2. This is not Ephrussi's terminology ; at the time he characterized the integration of tissues in culture as an intermediate step toward the integration of organisms. For a representative treatment, see, e.g., Ephrussi, 1935. Here are two key passages : *En résumé : dans la régénération des cultures nous avons une véritable régulation de la forme, telle que nous la connaissons dans des organismes complexes ; et nous arrivons à interpréter ce phénomène si nous admettons la subordination des éléments à l'activité de l'ensemble, subordination qui se ramène d'ailleurs, dans le cas précis, à l'action de facteurs relativement simples* (16-17). ... *Les cellules, toutes potentiellement identiques, équivalentes, se comportent comme si elles étaient différentes et ces différences sont uniquement fonction de leur situation, avec tout ce qui en découle. Mais aussi elles restent potentiellement équivalentes ; toutes gardent la potentialité totale. Nous avons donc là un début de différenciation étant uniquement fonction de situation, ne comportant ni division différentielle, ni restriction des potentialités* (22).

to yeast cells, Ephrussi exploited the *petites colonies* as part of a systematic exploration of nucleo-cytoplasmic relations (ref). This work was an important step toward the founding of mitochondrial genetics. Ephrussi himself did not participate directly in the formation of that sub discipline, although one of his closest co-workers, Piotr Slonimski, was one of the pioneers in that field. (An important early report is Ephrussi, 1949. A good review is available in Ephrussi, 1953).

5. *Ca.* 1960-1979. Returning to tissue culture, Ephrussi pioneered hybridization of somatic cells, helping to round somatic cell genetics, a field that developed possibilities that he had actively sought ever since the 1930s[3]. (This part of Ephrussi's career is reviewed in Burian, *et al.*, 1991, Weiss, 1992 ; see also Zallen and Burian, 1992).

I will say a bit more about this program of research below, reserving a close examination for another paper. The concern here is to examine the unifying themes and approaches that underlie this very diverse body of work.

II. The problems Ephrussi tackled exhibit considerable thematic unity and his experimental approach manifests considerable methodological unity even though the actual techniques he employed were quite varied. Though the point cannot be explored here, the unifying elements in his work can help us to understand recent developments in developmental genetics. As I reconstruct his work, the Ariadne's thread is his search for units of development, by which I mean a search for something for developmental biology comparable to units of selection (replicators, interactors, units of inheritance) in evolutionary biology and genetics[4]. *This search for units of development* characterizes all of his programs of research, with the possible exception of the very first. Although he never used such a label, his projects aimed to specify, localize, and understand the action of such units.

3. See, e.g., Ephrussi, 1953 : " Obviously, what is required is more than deductions from the behaviour of germ cells ; what is needed is direct genetic analysis of somatic cells, for the assumed functional equivalence of irreversibly differentiated somatic cells, however plausible, is only an hypothesis. Crosses between such cells being impossible, only nuclear transplantation from one somatic cell to another, or grafting of fragments of cytoplasm could provide the required information ; such experiments however will have to await the development of adequate technical devices.* In the meantime, the closest approximation to the evidence we would like to have is provided by the study of lower forms which propagate by vegetative reproduction and possess no isolated germ line. ...You will see that these studies confirm the view that the cytoplasm like the genes, is endowed with genetic continuity. The genes are therefore no longer to be regarded as the sole cell-constituents endowed with this property " (5-6). [*Ephrussi's footnote mentioned nuclear transplant experiments of Comandon and Fonbrune (Comandon and de Fonbrune, 1939) and Lorch and Danielli (Lorch and Danielli, 1950) on amoebae, and Briggs and King on frogs (Briggs and King, 1952). I add that Ephrussi had assigned nuclear transplantation experiments in frogs to Slonimski for his dissertation, but that they were unable to overcome the technical problems involved. I have been unable to relocate an earlier comment by Ephrussi, *ca.* 1939, suggesting that hybridization of somatic cells, if only feasible, would be the ideal way to study differentiation.]

4. My thinking on this topic has been significantly influenced by the excellent dissertation of Manfred Laubichler (Laubichler, 1997), which greatly clarifies the conceptual requirements for identifying functional units of various sorts and at various levels in biology.

Very early on, he recognized that there could be no strictly Mendelian account of the mechanisms by which commitment or determination of cells takes place in multicellular organisms. The point rests on the familiar paradox that if (virtually) all cells of an organism contain the same genes, there must be some additional mechanism, rigorously controlled, that determines in an orderly fashion which cells become cells of a given cell type. There are texts proclaiming this dilemma at every stage of his career. (See the epigraph above, from a 1962 text that summarizes the argument). Yet he sought a post-doctoral fellowship to spend 1934 in T.H. Morgan's laboratory because he also recognized the correctness of the fundamental account of reproduction and inheritance implicit in Mendelian genetics and wanted to employ genetic tools in his quest. This stance was decisively reinforced in 1934, when he showed that undifferentiated cells taken from a brachyuric mouse embryo, fated to die before it could produce cartilage, were already committed to form that tissue and could be made to do so. He interpreted this result as showing the fundamental importance of nuclear controls of cell determination. Certainly by this time, but probably already by 1930, *he* was committed too. *His* commitment was to finding whatever it is that is necessary to account for determination, differentiation, and integration of the organism. This meant both including *and going beyond* genetic mechanisms. The material substrate for the " whatever it is " that goes beyond genetic mechanisms is what I am characterizing as units of development. Among the candidates for consideration, over and above Mendelian genes, are the polarities, the various constituents, the stable states, and the organized structures of cells and embryos. The orderly commitment of cells in a given lineage to a particular fate and their subsequent differentiation cannot be understood except by reference to some or all of these factors. In the remainder of this brief paper I will explore the strategies and tactics of his search.

From his very earliest research, Ephrussi developed tactics for distinguishing between distinct mechanisms involved in bringing about a common effect. He became expert in finding systems of double control and disentangling them. Differentiation is precisely such an effect — it is achieved via a complex system involving (at least) double control. One system of control (which may itself be double) is the system that yields *determination* (that is, the commitment, often covert, of a cell to a particular fate) of cells, a second, perhaps also double, is that which triggers *differentiation,* as distinct from determination. For example, in his early work with tissue culture, he argued repeatedly against certain views of Christian Champy (see, e.g., Champy, 1912 and Ephrussi, 1931). Champy, a pioneer of tissue culture in France, maintained that fibroblastic cells in culture are indifferent because they have dedifferentiated to the point that they are no longer determined. Two of Ephrussi's arguments to the contrary are typical. (1) In specific conditions the cells in question (osteoblasts and fibroblasts from various sources) can be shown to manifest different behaviors, such as distinctive residual growths, according to their provenance, thus indicating that they have not lost all distinctive peculiarities (e.g., Ephrussi, 1931, pp. 19 ff.). (2) In ideal conditions, one can bring about differentiation of some of the cells in tissue culture. Cells from different sources differentiate (sometimes only partially) in precisely the ways that their parent cells did. Thus, the cells are not indifferent, but carry with them the commitment that Champy thought they had lost. They still contain some covert factor that commits them to one pathway rather than another.

The strategy here is maintained throughout Ephrussi's career : 1) Start with some seemingly unitary phenomenon — in this instance, phenotypic dedifferentiation of cells in tissue culture. 2) Seek to discriminate among the relevant mechanisms and/or structures underlying that phenomenon. 3) Entertain a wide range of hypotheses about those mechanisms and structures, seeking to find at least some that can be pairwise distinguished — or whose causal relevance to the phenomena can be distinguished — by what Ephrussi occasionally calls *données analytiques* (e.g., Ephrussi, 1933, p. 3). This means that the analytically relevant experimental findings make one of the mechanisms far more probable than the other. 4) Use the results to distinguish the roles of the processes or structures from one another or to get at the constituents or mechanisms involved, perhaps by elucidating some necessary conditions that must be met. (In redifferentiating cells, the experimental findings show that one relevant necessary condition is that the cells have maintained their specific determination by a covert cellular component, mechanism, or structure. Even without knowing what the mechanism is, Champy's account can be ruled out).

This sketchy description Ephrussi's research methodology is extremely brief, but it is all there is room for here. In general, Ephrussi draws on the findings of embryology, tissue culture, and other work to support such claims as, for example, that determination of nearly all cells is essentially irreversible. He uses such findings to argue against hypothetical models, both his own and others. For example, he uses the claim of irreversibility against Delbrück's models for determining cell commitment by means of dynamic equilibria. As of the mid-1940s to the mid 1950s, alternative models based on self-reproducing units in the cytoplasm are far more probable[5]. Although these models had to

5. " [Turning to the interpretation of differentiation as due to some sort of cytoplasmic elements], I wish first to repeat that whatever mode of cytoplasmic variation is postulated, it must be based on the assumption of self reproducing units. This is what Delbrück (1948) [*sic* : Delbrück, 1949], in a very alluring recent suggestion, has attempted to avoid. Instead, he resorts to the notion of flux equilibrium between mutually exclusive reactions, capable of shifting the cell under the influence of transitory changes of the environment from one state of stable equilibrium into another, alternative one. Aware, no doubt, of the irreversible character of certain cell differentiations, he emphasizes that systems of this sort would be capable of all possible degrees of stability, and that the shifts from one stable state to another could be either reversible or irreversible. I feel however that as soon as the occurrence of irreversible change is assumed, there is no real escape from the conclusion that a self reproducing unit or a part of it is involved " (Ephrussi, 1951, 243-244. See also Ephrussi, 1949, 179-180).

be replaced during the decade that followed, Ephrussi here encapsulates the heart of the argument for the existence of plasmagenes or some sort of similar entities.

At the same time, his use of *all* such models is always epistemologically reserved and cautious. There can be no question of straightforward proof by arguments like this. Ephrussi's method, rather, was to push the models ever further until they either collapsed or were replaced by better models. This attitude was perhaps carried to an extreme, as is illustrated by his great reluctance to affirm that the particles responsible for the respiratory deficiency in yeast are mitochondria. He did not make that claim until about five years after it was generally accepted. Similarly, until very late in the game he did not resolve the question whether the hypothetical particles had been lost from the cell or had ceased to function due to some sort of block. The epistemological standard that had to be met in order to establish the existence of an indispensable component or mechanism was very high, and the *données analytiques* had to cover all of the alternative possibilities.

III. To conclude, I offer some remarks on the ambiguity of the hypotheses that Ephrussi explored and the epistemological difficulty of establishing any of them experimentally. There were always at least two possibilities for finding units of development : finding *constituents* of cells that did not enter into the genetic account of the capacities of cells on the one hand, and, on the other, finding units of *structure* or *function* that provided the conditions required for the function or activation of independently identified genetic units. One difficulty in completing the search for units of development along the lines explored by Ephrussi is that it is often extremely hard to disentangle these alternative possibilities — especially since the units in question may be evanescent, may act only during very brief critical periods, and/or may be extremely hard to detect. (After all, determination takes place very early, the entities or mechanisms responsible for determination may well be consumed in the process, and there may be no straightforward means of registering their presence).

To revert to the example of mitochondria in *petite colonie* yeast for a simple illustration of the point, there are at least five possibilities to explore in determining whether some sort of mitochondrial deficiency is responsible for the respiratory incompetence of the yeast cells. (1) The mitochondria have been lost and all that are left are mitochondrial ghosts. (2) There has been some sort of structural change in the mitochondria blocking their function — a change which is highly specific and which could be reversed if one knew exactly what it were. (3) Some specific component of the mitochondria has been irreversibly lost. (4) The responsible unit is some other particulate component of the cell, not visualizable by current techniques. And (5) the phenomenon is the consequence of some combination of nuclear genes, perhaps combined into something like a supergene, affecting the relevant cytoplasmic particles. I mentioned earlier that Ephrussi hesitated enormously in identifying the particles responsible for the deficiency as mitochondria and that he did not take part in the development of mitochondrial genetics. Herschel Roman offers a somewhat harsh, but nonetheless largely just, judgement about this : It is a curious fact that even in the later papers [on yeast] by Ephrussi and his collaborators, no mention is made of mtDNA as a possible site of the petite mutation. Although he could not have been oblivious of the fact that mitochondria contain DNA or that a relationship had been established between mtDNA and petiteness..., Ephrussi seemingly was reluctant to give a physical reality to his cytoplasmic factor. Not much is gained by speculations on the reasons for this omission, but two possible explanations come to mind as being worthy of consideration. First, Ephrussi may have lost his earlier enthusiasm for cytoplasmic inheritance when he shifted his considerable talents to somatic cell genetics in mice. Second, and just as likely, he was trained in experimental embryology and took pride in his background as a biologist. He may have preferred to retain a certain ambiguity about his cytoplasmic factor rather than hazarding guesses as to its molecular nature (Roman, 1982, p. 3-4).

To this I will add two other possibilities. (1) The harsh scepticism required by Ephrussi's method, made habitual, probably played some role in Ephrussi's personal choices regarding commitment to *any* model. (2) With his entry into somatic cell genetics, Ephrussi may have finally had to give up on the idea that there were definitive biological *entities* of a non-genetic sort, as opposed to units of organisation, peculiar to development. I hope to take up this and related issues in an expanded version of this paper that will focus especially on the last phase of Ephrussi's career and on its methodological interest for contemporary developmental biology. The general point put forward here, however, remains of interest to his successors in the new biology of development. It will always remain enormously difficult to balance the tendency toward excessive credulity in constructing and supporting hypotheses about entities as complex and hidden as the putative units of development sought by Ephrussi and the scepticism with which those hypotheses must be treating in submitting them to the rigorous test by means of *données analytiques*. That remains an ongoing task that must be faced in the new biology of development.

ACKNOWLEDGEMENTS

The idea of units of development arose as the indirect product of reading the excellent dissertation of Manfred Laubichler (Laubichler, 1997), which greatly clarifies the conceptual requirements for identifying functional units of various sorts and at various levels in biology. Much of the material in this paper was developed in the course of a long term collaboration with Jean Gayon and Doris Zallen. I am particularly grateful to them and to Piotr Sionimski and Mary Weiss for information provided in a series of interviews.

REFERENCES

Robert W. Briggs and J. King Thomas, " Transplantation of Living Nuclei from Blastula Cells into Enucleated Frogs Eggs ", *Proceedings of the National Academy of Science,* 38 (1952), 455-463.

Richard M. Burian, " La contribution française aux instruments de recherche dans le domaine de la génétique moléculaire ", in Jean-Louis Fischer and William H. Schneider (eds), *Histoire de la génétique : Pratiques, techniques et theories*, Paris, ARPEM, 1990, 247-269.

Richard M. Burian and Jean Gayon, " Genetics after World War II : the laboratories at *Gif* ", *Cahiers pour l'histoire du CNRS ,* 7 (1990), 25-48.

Richard M. Burian and Jean Gayon, " Génétique et recherche médicale en France : le cas de Boris Ephrussi (1901-1979) ", *Sciences Sociales et Santé,* 10 (1992), 25-45.

Richard M. Burian, Jean Gayon and Doris Zallen, " The Singular Fate of Genetics in the History of French Biology, 1900-1940 ", *Journal of the History of Biology,* 21 (1988), 357-402.

Richard M. Burian, Jean Gayon and Doris Zallen, " Boris Ephrussi and the synthesis of genetics and embryology ", in Scott Gilbert (ed.), *A Conceptual History of Embryology*, New York, Plenum, 1991, 207-227.

Richard M. Burian and Doris T. Zallen, " The non-interaction of regulatory genetics and human cytogenetics in France, 1955-1975 ", in Krishna R. Dronamraju (ed.), *The History & Development of Human Genetics*, Singapore, World Scientific, 1992, 92-101.

Christian Champy, " Sur les phénomènes cytologiques qui s'observent dans les tissus cultivés en dehors de l'organisme ", *Comptes rendus de la Société de Biologie,* 72 (1912), 987-988.

J. Comandon and P. de Fonbrune, " Greffe nucléaire totale, simple ou multiple, chez une *Amibe "*, *Comptes rendus de la Société de Biologie,* 130 (1939), 744-748.

M. Delbrück, " [Discussion of] Influences des gènes, plasmagènes, et du milieu dans le déterminisme des charactères antigéniques chez *Paramecium aurelia* (variété 4) by T. Sonneborn and G.H. Beale ", *Editions du Centre National de la Recherche Scientifique,* Paris, 1949, 33-35.

Boris Ephrussi, " Résultats recents de la culture des tissus ", *Annales et Bulletin de la Société Royale des Sciences Médicales et Naturelles de Bruxelles,* n° 7-8 (1931), 15-44.

Boris Ephrussi, " Contribution à l'analyse des premiers stades du développement de l'oeuf. Action de la température ", *Archives de Biologie,* 44 (1933), 1-147 + 1 planche.

Boris Ephrussi, " Sur le facteur lethal des Souris brachyures ", *Comptes rendus de l'Académie des Sciences,* 197 (1933), 96-98.

Boris Ephrussi, " The behavior in vitro of tissues from lethal embryos ", *Journal of Experimental Zoology,* 70 (1935), 197-204.

Boris Ephrussi, *Phénomènes d'intégration dans la culture des tissus,* Edited by Emmanuel Fauré-Frémiet, Vol. IV, *Exposés de Biologie (Embryologie et Histogenèse),* Paris, Hermann, 1935.

Boris Ephrussi, " Analysis of Eye Color Differentiation in Drosophila ", *Cold Spring Harbor Symposia on Quantitative Biology,* 10 (1942), 40-48.

Boris Ephrussi, " Chemistry of 'Eye Color Hormones' of Drosophila ", *Quarterly Review of Biology,* 17 (1942), 327-338.

Boris Ephrussi, " Action de l'acriflavine sur les Levures ", *Unités biologiques douées de continuité génétique,* Paris, Centre National de la Recherche Scientifique, 1949, 165-180.

Boris Ephrussi, " Remarks on Cell Heredity ", in L.C. Dunn (ed.), *Genetics in the Twentieth Century,* New York, Macmillan, 1951, 241-262.

Boris Ephrussi, *Nucleo-cytoplasmic Relations in Micro-organisms. Their Bearing on Cell Heredity and Differentiation,* Oxford, England, Oxford University Press, 1953.

Boris Ephrussi, " Mendelism and the New Genetics ", *Somatic Cell Genetics,* 5 (1979), 681-695.

J.R.S. Fincham, " Boris Ephrussi - his work in genetics ", *BioEssays,* 14 (1992), 347-348.

Jean Gayon, " Génétique de la pigmentation de l'oeil de drosophile : la contribution spécifique de Boris Ephrussi ", in Claude Debru, Jean Gayon and Jean-François Picard (eds), *Les sciences biologiques et médicales en France, 1920-1950,* Paris, CNRS Éditions, 1994, 9-23.

Scott F. Gilbert, " Epigenetic landscaping : C.H. Waddington's use of cell fate bifurcation diagrams ", *Biology and Philosophy,* 6 (1991), 135-154.

Manfred Laubichler, *Identifying Units of Selection : Conceptual and Methodological Issues,* Ph. D., Yale University, 1997.

I.J. Lorch and J.F. Danielli, " Transplantation of Nuclei from Cell to Cell ", *Nature,* 166 (1950), 329-330.

A. Lwoff, " Recollections of Boris Ephrussi ", *Somatic Cell Genetics,* 5 (1979), 677-679.

Herschel Roman, " Boris Ephrussi ", *Annual Review of Genetics,* 14 (1980), 447-450.

Herschel Roman, " Boris Ephrussi and the Early Days of Cytoplasmic Inheritance in Saccharomyces ", in Piotr P. Slonimski, Piet Borst and Giuseppe Attardi (eds), *Mitochondrial Genes,* Cold Spring Harbor, N. Y., Cold Spring Harbor Laboratory, 1982, 1-4.

Herschel Roman, " The Early Days of Yeast Genetics : A Personal Narrative ", *Annual Review of Genetics,* 20 (1986), 1-12.

Jan Sapp, *Beyond the Gene : Cytoplasmic Inheritance and the Struggle for Authority in Genetics,* New York, Oxford University Press, 1986.

Jan Sapp, " Inside the Cell : Genetic Methodology and the Case of the Cytoplasm ", in J.A. Schuster and R.R. Yeo (eds), *The Politics and Rhetoric of Scientific Method*, Dordrecht, D. Reidel, 1986, 167-202.

Mary C. Weiss, " Contributions of Boris Ephrussi to the development of somatic cell genetics ", *BioEssays,* 14 (1992), 349-353.

Doris T. Zallen and Richard M. Burian, " On the beginnings of somatic cell hybridization : Boris Ephrussi and chromosome transplantation ", *Genetics,* 132 (1992), 1-8.

EPHESTIA : THE EXPERIMENTAL DESIGN OF ALFRED KÜHN'S PHYSIOLOGICAL GENETICS

Hans-Jörg RHEINBERGER

This paper aims at reconstructing some steps in the experimental pathway that led Alfred Kühn and his collaborators, in particular Karl Henke, Ernst Caspari, Ernst Plagge and Erich Becker to their specific form of developmental physiological genetics. Special attention will be paid to the interaction of the different elements, rooted in different techniques and research traditions, of the Ephestia experimental system. It will be shown how they came to form a mixed experimental set-up composed of genetic, embryological, physiological, and biochemical constituents that helped to conceive of " gene action chains " as acting on " substrate chains ".

Alfred Kühn had already made his career as a distinguished zoologist in Freiburg, Berlin, and Göttingen when he decided, in 1924, to settle on the moth *Ephestia* kühniella Zeller as a model organism to study what he called " developmental physiological genetics ". Between 1924 and 1929, Kühn and Karl Henke focused on an analysis of the wing pattern of Ephestia by means of selecting mutants and subjecting them to a systematic variation of external developmental stimuli.

In 1929, Kühn and Henke found a red-eyed moth in their cultures that followed a Mendelian recessive inheritance pattern. The pleiotropic nature of the eye mutation that extended to the coloration of the testicles and the body color of the caterpillar of Ephestia induced Kühn's doctoral student Ernst Carspari in 1931 to think of an alternative combination of experimental techniques. He decided to look at the effect of implanting testicles from mutant flies into wild type flies and vice versa. What he observed was that wild type testicles were able to darken mutant fly eyes, and that mutant testicles became colored when implanted into wild type caterpillars. The conclusion was at hand that the wild

type organs secreted a substance into the hemolymph that induced the formation of pigment, and that this substance was lacking in the mutant[1].

This was an experimental track to follow for Caspari, Kühn, and other co-
workers who soon came to join the Göttingen team. Within a few years, and
with the financial help of the Rockefeller Foundation, the group restructured
their moth work around the transplantation system. The secreted substance
even appeared to diffuse into the eggs of mutant flies, causing the eyes of the
mutant offspring to become pigmented. The conclusion was at hand that the
substance acted as a hormone. Another crucial observation was that the amount
of substance released from homozygous and heterozygous implants was the
same, but that the effect of the implantation of two such organs was additive.
This induced Kühn, Caspari and Ernst Plagge to differentiate between a
" primary reaction in the plasm " of cells carrying the wild type gene that
accounted for the effect of dominance, and the subsequent production of the
presumed hormone that in turn caused pigmentation, resulting in a series of
" intermediates in the reaction sequences that lead from a particular gene to the
mature characters of organization "[2].

How to go about the nature of the hormone-like active substance ? It
became obvious at this point to Kühn and his co-workers that their experimental system needed to be expanded by a biochemical in vitro component. Meanwhile, Boris Ephrussi and George Beadle had shown via transplantation and
the injection of extracts that the substances correlated with the v and cn factors
in Drosophila eye pigment formation complemented each other in a reaction
chain[3]. Concomitant with his move from the University of Göttingen to the
Kaiser Wilhelm Institute for Biology in Berlin-Dahlem in 1937, Kühn
recruited and mobilized biochemical skills. In that same year, Erich Becker,
who had come with Kühn from Göttingen and Plagge showed in parallel
extraction and injection experiments that the active substances in Drosophila
and Ephestia were exchangeable and thus equivalent.

Erich Becker joined forces with Wolfhard Weidel and Adolf Butenandt from
the Kaiser Wilhelm Institute for Biochemistry, Butenandt had come from Dan-
zig to Berlin Late in 1936. It took them two and a half more years in an ever
more competitive race with George Beadle in Stanford to identify their pre

1. Ernst Caspari, " Über die Wirking eines pleiotropen Gens bei der Mehlmotte Ephestia kühniella Zeller ", *W. Roux' Archiv für Entwicklungsmechanik der Organismen,* 130 (1933), 353-381.

2. Alfred Kühn, Ernst Caspari and Ernst Plagge, " Über hormonale Genwirkungen bei Ephestia
kühniella Z ", *Nachrichten von der Gesellschaft der Wissenschaften zu Göttingen, Mathematisch-
Physikalische Klasse,* (1935), 1-29.

3. On the work of Ephrussi and Beadle see Richard M. Burian, Jean Gayon and Doris Zallen,
" The singular fate of genetics in the history of French biology, 1900-1940 ", *Journal of the History of Biology,* 21 (1988), 357-402.

sumed hormone as kynurenine, a derivative of the amino acid tryptophane. The structure was published in January 1940[4].

The hormone hypothesis that had led Kühn and his working group into biochemistry had there at the entrance to be buried. A major rearrangement resulted from grafting a biochemical branch onto the Ephestia system. A decisive new feature emerged : Kühn started to distinguish between what he now called a " gene action chain " on the one hand, and a " substrate chain " on the other[5]. The earlier, linear " reaction sequence with intermediates " had become two-dimensional. The gene reaction chain resulted in " ferments " which in turn intervened as enzymes in a metabolic chain of transformations leading from the amino acid tryptophane to the pigments of the insect eye. The intercalation of both chains visualizes the experimental integration of physiology and genetics into a " physiological genetics " that had come to a molecular resolution with respect to one of its dimensions, the substrate chain.

Kühn concluded his 1941 paper by stating : " We stand only at the beginning of a vast research domain. ...Our apprehension of the expression of hereditary traits changes from a more or less static and preformistic conception to a dynamic and epigenetic one. ...Every step in the realization of characters is, so to speak, a knot in a network of reaction chains into which many gene actions irradiate. ...Only a methodically conducted genetic, developmental and physiological analysis of a great number of single mutations can gradually disclose the Wirkgetriebe der Erbanlagen "[6].

    4. Adolf Butenandt, Wolfhard Weidel and Erich Becker, " Kynurenin als Augenpigmentbildung auslösendes Agens bei Insekten ", *Die Naturwissenschaften,* 28 (1940), 63-64.
    5. Alfred Kühn, " Über eine Gen-Wirkkehtte der Pigmentbildung bei Insekten ", *Nachrichten der Akademie der Wissenschaften in Göttingen, Mathematisch-Physikalische Klasse,* (1941), 231-261.
    6. Alfred Kühn, *art. cit.,* 258.

LEVI-MONTALCINI ET LA DÉCOUVERTE DU FACTEUR DE CROISSANCE DU NERF

Jean-Claude DUPONT

Dans sa *Nobel Lecture* (1986)[1] Levi-Montalcini reprend à son compte un jugement particulièrement dur de P. Medawar sur l'embryologie expérimentale des années quarante et dans son livre autobiographique[2] elle insiste sur le climat pessimiste ou " antidéterministe " qui prévalait dans cette période qui précéda sa propre découverte du facteur de croissance. Ce climat contrastait selon elle avec l'enthousiasme qui avait suivi les grands travaux embryologiques des années trente. On a tenté de comprendre cette analyse sévère de Levi-Montalcini et de réfléchir sur les bases historiques véritables de sa découverte. On peut suivre le cheminement expérimental de la neurobiologiste italienne au cours de ses pérégrinations géographiques : Turin (1932-1947), St Louis (1947-1952), Rio (1952-1953), de nouveau St Louis (1953-1961), et enfin Rome (1961-1979). D'une part Levi-Montalcini se défend d'avoir jamais eu un programme de recherche ou une stratégie à long terme. Elle évoque une série d'observations fortuites : ainsi ses observations de la mort neuronale au niveau médullaire en 1947. Ailleurs ce sont des hasards expérimentaux comme la découverte du venin de serpent comme source de NGF. D'autre part elle dit réinterpréter certains résultats d'autres chercheurs (expériences de Hamburger, de Bueker) systématiquement dans le sens précis d'une activité neurotrophique, certains de ses propres résultats impliquant même une faculté " d'oubli de l'information négative "[3]. S'il paraît difficile de nier le rôle du hasard dans les travaux et les résultats de Levi-Montalcini, il apparaît que ses interprétations,

1. R. Levi-Montalcini, " The nerve growth factor : thirty-five years later ", in T. Frängsmyr and J. Lindsten (eds), *Nobel Lectures in Physiology or Medicine (1981-1990)*, London ; River Edge, World Scientific Publishing Co., 1993, 346-370.
2. R. Levi-Montalcini, *Éloge de l'imperfection*, Paris, Plon, 1988, 133.
3. La " loi de l'oubli de l'information négative " de A.R. Luria (psychologue soviétique) est rappelée par Levi-Montalcini (*Ibid.*, p. 163) : - Tous les faits qui vont dans le sens de l'hypothèse formulée sont immédiatement acceptés et mis en évidence ; - Tous les faits contraires sont considérés comme insignifiants et sont oubliés.

réinterprétations et oublis ne relèvent pas d'une volonté de rupture avec le climat scientifique de l'époque ainsi qu'elle le laisse entendre. Car la problématique est déjà bien posée. Les travaux de Levi-Montalcini correspondent à un des soucis majeurs de la neuroembryologie du moment, à savoir l'étude des interactions entre populations de neurones périphériques et leurs organes-cibles. Il s'agissait déjà de comprendre comment les fibres nerveuses se projettent au cours du développement pour établir des connexions avec leurs organes-cibles respectifs, d'étudier la sensibilité des populations cellulaires du système nerveux aux influences exercées par le champ d'innervation périphérique. Les populations neuronales, leurs connexions et le parcours des fibres nerveuses étaient-elles déterminées par la nature ou les propriétés des organes-cibles ou par un code génétique extrêmement précis ? L'enjeu était de comprendre l'interaction des facteurs génétiques et épigénétiques dans le contrôle des processus de différenciation du système nerveux dans les premières phases de son développement. Les controverses de Weiss et de Sperry qui ont marqué cette époque et auxquelles Levi-Montalcini assistera, seront révélatrices de cette préoccupation majeure de la neuroembryologie naissante.

Les études de Weiss sur la morphogenèse, ses travaux sur les mécanismes de migration cellulaires, sur la migration axonale, sur les effets du contact sur l'orientation des migrations cellulaires visent à comprendre la matérialité du concept de champ sans faire appel à un quelconque principe entéléchique (Driesch). La possibilité pour le neurone d'adapter sa morphogenèse à des conditions expérimentales nouvelles, c'est à dire à un environnement différent, pouvait impliquer une plasticité du neurone qui ne serait pas alors soumis à un déterminisme génétique strict, qui du reste était à l'époque incompréhensible, mais surtout à des facteurs extrinsèques. Admettre un principe d'équipotentialité pour le neurone pouvait ainsi impliquer un rejet de la spécificité neuronale. Comme le souligne Levi-Montalcini, les thèses embryologiques de Weiss qui prédominent à l'époque sont compatibles avec la réactualisation des thèses antilocalisationnistes de Lashley et de Goldstein et des théoriciens de la *gestalt* qui font aussi l'objet d'un consensus assez large, en dépit de données cliniques ou expérimentales divergentes. Notons que d'un point de vue histologique, le réticularisme encore très actif, en excluant une théorie des localisations cérébrales strictes, sous-estimait la spécificité des connexions neuroniques (et du coup, surestimait le rôle de la plasticité dans le développement) et était favorable au principe d'équipotentialité. En fait, la neuroembryologie piétine et ne peut qu'hésiter entre l'adoption pour le développement d'un modèle de déterminisme génétique ou épigénétique, ou si l'on préfère, elle est la traduction en ce qui concerne le système nerveux des relations problématiques entre épigénétisme et génétisme de l'embryologie naissante. Il n'y a pas dans les années quarante de construction possible d'une théorie du développement au sens d'une théorie mendélienne de l'hérédité, ce qui donne à l'embryologie une apparence de statisme et d'empirisme pur après les avances conceptuelles des années trente et les brillants travaux de Spemann. Certes on constate un effort des embryologistes vers une compréhension " moléculaire " ou biochimique des mécanismes de développement, mais ces tentatives restent le plus souvent infructueuses, témoins par exemple les échecs répétés d'isolement d'une substance biologique capable de reproduire l'induction neurale chez les amphibiens. Par ailleurs Prochiantz (1988) a raison de souligner que la notion de programme génétique est proche à l'époque d'un principe entéléchique. Il est donc exact que le contexte biochimique et génétique est peu favorable à la neuroembryologie naissante. Levi-Montalcini sera d'ailleurs vers 1947 tentée de rejoindre le groupe du phage en arguant de ses compétences dans le domaine des cultures cellulaires, vu le manque de perspective dans sa discipline qui lui semblait pour longtemps condamnée à rester descriptive. S. Luria, turinois d'origine comme elle, l'en dissuadera sans conviction.

Mais son pessimisme était sans doute excessif concernant le contexte histologique. Face aux difficultés théoriques d'ordre biochimique et génétique, la vieille histologie était resté longtemps l'outil de recherche privilégié et elle devenait de plus en plus expérimentale. Lorsqu'au début des années trente, l'étudiante Levi-Montalcini travaille à Turin sous la direction de Levi, les histologistes italiens sont encore partagés entre réticularistes (Golgi) et neuronistes comme Levi et Lugaro. La maîtrise de la technique d'imprégnation chromo-argentique (Golgi, 1875) avait permis l'identification de milliers de populations neuronales du système nerveux. De plus la stratégie des neuronistes comme His et Cajal avait très tôt consisté en l'analyse du système nerveux en voie de développement au stade ou celui-ci ne comporte que quelques milliers de cellules sur le point d'entrée en différenciation, de manière à consolider la théorie du neurone, et venir ainsi à bout des objections réticularistes. L'histogenèse du système nerveux avait donc d'emblée été approfondie, en partie parce qu'il s'agissait d'étendre la théorie cellulaire à tout l'organisme. Elle aboutira à partir des travaux de His à concevoir le développement, conformément aux vues de Roux pour l'embryologie, en terme de mécanique de migrations, différenciations et prolifération cellulaires et donnera lieu à des controverses sur le lignage des cellules nerveuses, la formation de la crête neurale et de ses dérivés. Or la neurohistologie avait largement permis d'aborder la problématique " génétisme-épigénétisme ". Car un des problèmes de fond de cette histogenèse du système nerveux était celui du nombre de cellules nerveuses de structures identifiables, qui pourrait être fixe ou soumis aux fluctuations des facteurs ambiants comme en témoignent les premiers travaux neurohistologiques de Levi-Montalcini, consistant en des comptages pénibles dont elle se lassera vite. Mais elle apprendra les délicates techniques de colorations et de réalisation des coupes histologiques. Elle s'initiera aussi aux nouvelles techniques de cultures cellulaires (tissus conjonctifs, musculaires, nerveux…) et à l'étude des cellules animales in vitro auprès de Herta Mayer qu'elle retrouvera ensuite au Brésil. Elles furent importées en Italie par Levi dès 1928 qui avait réalisé leur application possible à l'analyse des capacités de prolifération et de différenciation des cellules des différents types dans les milieux et conditions expérimentales différents. Levi-Montalcini étudiera ainsi pour sa thèse la formation des réseaux de collagène dans les tissus conjonctifs, musculaires et épithéliaux. Ce sont ces techniques qu'elle réutilisera vingt ans plus tard et qui permettront la découverte du NGF. C'est aussi chez Levi qu'elle apprend les manipulations de l'embryon de poulet, matériel biologique facile à se procurer. Elle y réalisera même en 1938, en collaboration avec le neurologue Visintini, une étude de la différenciation des centres et circuits neuronaux qui se manifestent dans la motilité et la sensibilité de l'embryon grâce à l'enregistrement oscillographique de l'activité spontanée et provoquée par des électrodes stimulatrices du cerveau des embryons de poulet.

Les travaux de Levi-Montalcini et de Hamburger s'inscrivent donc dans la tradition d'une neurohistologie qui n'est plus seulement restée descriptive mais résolument tournée vers l'expérimentation, commencée au début du siècle avec Harrison. Elle se développera après les travaux pionniers de Braus, Shorey, Weiss, Detwiller, avec les nombreuses expériences visant à modifier le champ périphérique d'innervation au cours du développement au moyen de sections ou de greffes de membre, expériences suivies d'examen histologique et/ou d'études comportementales. Harrison tentera ainsi les xénogreffes et Hamburger des greffes d'ébauches de membres sur des embryons de poulet et des amputations, confirmant l'influence du champ périphérique sur le système nerveux embryonnaire. Influencé par Spemann, il conclura à l'absence après l'amputation d'un facteur d'induction normalement libéré par les tissus embryonnaires pendant les stades précoces du développement, thèse que reprendra son élève Bueker, avant de se rallier à l'idée d'un facteur trophique. Le problème est que l'interprétation des résultats histologiques est difficile surtout après les manipulations chirurgicales de l'embryon. Ainsi l'examen après amputation ou greffe révèle hypoplasies et hyperplasies neuronales respectivement ce qui est interprété longtemps comme une preuve de la régulation de la prolifération des populations neuronales par les tissus innervés. Pour Hamburger et Levi-Montalcini les facteurs périphériques permettent la prolifération et la différenciation initiale des cellules indifférenciées non encore connectées, et la croissance et la survie des neurones après la première croissance axonale. Plus tard Hamburger montrera que le nombre de motoneurones de l'embryon de poulet diminue bien à la suite d'une augmentation de la mort neuronale et non d'un défaut de prolifération et de différenciation (1958). D'après Jacobson (1991) le NGF sera compris comme ayant des propriétés mitogènes longtemps après sa découverte à cause de cette méprise initiale sur son activité. Les débat sur son activité biologique tiennent aussi à ce que même si l'on refusait la théorie de Haeckel selon laquelle tout processus régressif faisait partie d'une récapitulation phylogénétique de l'ontogénie, la mort neuronale anciennement décrite par Beard (1896), puis Ernst (1926) et que Levi-Montalcini avait étudiée de manière détaillée au niveau de la moelle épinière, pouvait dans une optique déterministe difficilement être conçue comme partie intégrante d'un processus ontogénétique normal. Ce n'est qu'à partir des années soixante-dix que la croissance du système nerveux sera pensée en terme de programme de prolifération, migration et différentiation et mort cellulaire. La survie de tout neurone dans l'embryon semble dépendre, en réalité, d'une combinaison de facteurs. Les facteurs neurotrophiques et les autres agents de l'environnement semblent avoir pour fonction de réprimer un " programme de suicide " qui autrement s'exprimerait normalement. L'histoire des facteurs de croissance ne fera que commencer après la découverte du NGF.

Si le hasard n'est jamais absent, il apparaît que, percevant la problématique embryologique de l'époque, Levi-Montalcini a su combiner des approches expérimentales à la pratique de certaines desquelles elle s'était préparée pendant de longues années : histologie avec Levi, culture cellulaire avec Meyer, embryologie expérimentale avec Hamburger et biochimie avec Cohen, ceci pour initier dans les années cinquante l'histoire des neurotrophines. Davantage qu'avoir rompu avec un contexte prétendument antidéterministe, il apparaît qu'elle a contribué à donner à la neuroembryologie, par la matérialisation du concept de champ, le statut de science moléculaire que la neurophysiologie par la matérialisation du concept de message nerveux, et que la génétique par la découverte de la nature chimique du gène, étaient en passe d'obtenir. En cela, sans doute assimile-t-elle trop rapidement matérialisme et déterminisme.

Références

E.D. Bueker, " Implantation of tumors in the hindlimb field of the embryonic chick and developmental response of the lumbosacral nervous system ", *Anat. Res.,* 102 (1948), 369-390.

C. Cohen, R. Levi-Montalcini and V. Hamburger, " A nerve growth stimulating factor isolated from sarcoma 37 and 180 ", *Proc. Natl. Acad. Sci.,* USA 40 (1954), 1014-1018.

S. Cohen, " Purification of a nerve growth promoting protein from the mouse salivary glands and its neurocytotoxic antiserum ", *Proc. Natl. Acad. Sci.,* USA 46 (1960), 302-311.

S.F. Gilbert, *Biologie du développement*, Paris, Bruxelles, De Boeck Université, 1996.

V. Hamburger, " The development and innervation of transplanted limb primordia of chick embryos ", *J. Exp. Zool.,* 80 (1939), 347-389.

V. Hamburger, " The effects of wing bud extirpation on the development of the central nervous system in chick embryos ", *J. Exp. Zool.,* 68 (1934), 449-494.

V. Hamburger, " Motor and sensory hyperplasia following limb-bud transplantations in chick embryos ", *Physiol. Zool.,* 12 (1939), 268-284.

V. Hamburger, " Regression versus peripheral control of differentiation in motor hypoplasia ", *Am. J. Anat.*, 102 (1958), 365-410.

V. Hamburger and R. Levi-Montalcini, " Proliferation, differentiation and degeneration in die spinal ganglia of the chick embryo under normal and experimental conditions ", *J. Exp. Zool.*, 111 (1949), 457-501.

M. Jacobson, *Developmental Neurobiology*, New York, London, Plenum Press, 1991.

R. Levi-Montalcini, " Effect of mouse tumor transplantation on the nervous system ", *Ann. N. Y. Acad. Sci.*, 55 (1952), 330-343.

R. Levi-Montalcini, *Éloge de l'imperfection*, Paris, Plon, 1988.

R. Levi-Montalcini, " The nerve growth factor : thirty-five years later ", in T. Frängsmyr and J. Lindsten (eds), *Nobel Lectures in Physiology or Medicine (1981-1990)*, London, River Edge, World Scientific Publishing Co., 1993, 340-370.

R. Levi-Montalcini, " The origin and development of the visceral system in the spinal cord of the chick embryo ", *J. Morphol.*, 86 (1950), 253-283.

R. Levi-Montalcini and P. Calissano, " Le facteur de croissance du nerf ", *Pour la Science*, 79 (1979), 12-22.

R. Levi-Montalcini and S. Cohen, " In vitro and in vivo effects of a nerve growth stimulating agent isolated from snake venum ", *Proc. Nat. Acad. Sci.*, USA 42 (1956), 695-699.

R. Levi-Montalcini and V. Hamburger, " Selective growth stimulating effect of mouse sarcoma on the sensory and sympathic nervous system of the chick embryo ", *J. Exp. Zool.*, 116 (1951), 321-362.

R. Levi-Montalcini and G. Levi, " Recherches quantitatives sur la marche du processus de différentiation des neurones dans les ganglions spinaux de poulet ", *Arch. Biol.*, (Liège) 54 (1943), 183-206.

R. Levi-Montalcini, H. Mayer, and V. Hamburger, " In vitro experiments on the effect of mouse Sarcoma 180 and 37 on the spinal and sympathetic ganglia of the chick embryo ", *Cancer Res.*, 14 (1954), 49-59.

A. Prochiantz, *Les stratégies de l'embryon*, Paris, PUF, 1988.

F. Visintini and R. Levi-Montalcini, " Relazione tra differenziazione strutturale e funzionale dei centri e dell vie nervose nell' embrione di pollo ", *Schweiz. Arch. Neurol. Neurochim. Psychiat.*, 43 (1939), 381-393.

P. Weiss, " Experiments on cell and axon orientation in vitro. The role of colloidal exsudates in tissue organisation ", *J. Exp. Zool.*, 100 (1945), 355-386.

P. Weiss, " In vitro experiments on the factors determining the course of the out-growing nerve fiber ", *J. Exp. Zool.*, 68 (1934), 393-448.

P. Weiss, *L'archipel scientifique*, Paris, Maloine, 1974.

The History of the Hemopoietic Stem Cell Concept

Patrick TRIADOU

The origin of the differentiated peripheral blood cells has been a central problem in haematology for a long time. Since haematologists were able to identify with microscopic examination different types of cells on blood and marrow films they rose up the question of their origin. During embryogenesis most tissues depend on continuous renewal of progenitor cells, but in adult life, only a few such as lymphoid tissue and bone marrow retain this characteristic. In the early decades of this century battle raged between great haematologists concerning the morphological identity of the hemopoietic stem cells and concerning the question whether they are pluripotential or unipotential. Morphological arguments could not settle this old problem which required new strategies and new experimental methodologies. Grafting experiments and later on in vitro culture systems allow the study of no recognizable precursor cells which correspond to what McCulloch called " a chapter of haematology without microscope ". With these data the original simple stem cell concept changed. Demonstrating the capacity of progenitor cell for extensive proliferation, differentiation and self-renewal, animal studies gave a functional definition of the stem cell which fulfils the theoretical criteria in the sense that it is capable of maintaining its own number during adult life in spite of cell loss. Convincing evidence has accumulated which suggest that there may be intermediate probably committed but morphologically unrecognized cell populations. Clonal culture systems showing that different types of *in vitro* colony forming cells exist with an absolute requirement for specific medium lead to give up the simple idea of a stem cell which on appropriate demand can differentiate into recognizable erythroid or granulocytic cell. Primitive progenitor cells have partially to be understood as a compartment of various committed stem cells with a continuum of properties and the question of differentiation discussed in function of colony stimulating factors. On the other hand monoclonal antibodies offer a powerful tool in order to characterize directly these unrecognized cells but do not succeed in founding specific markers. Therefore the myth of the stem cell becomes a reality without fulfilling the usual criteria of definition which include morphological and biochemical evidences.

We review in this paper the major advances constituting the indirect proof of the existence of the hemopoietic stem cell which can not be separated from the discovery of the different haematological growth factors. The latter participate in the biochemical understanding of the concept of micro environment.

First evidence for a pluripotent stem cell

When the concept has been developed, stem cells have been defined as morphologically unrecognized cells that have the role of providing blood differentiated cells during animal life span. Two alternatives could explain this property : extensive capacity of proliferation or the existence of a large store of stem cells at the onset of adult life. The second subject of debate was whether the stem cells are unipotential or multipotential. Each major cell line has its own stem cells or all the blood cells originate from one stem cell population (1). The question has been resolved by introducing transplantation experiments.

The cellular basis for hemopoietic repopulation was established in early 1950s in the work of Ford on lethally irradiated mice (2). A possible assay for repopulating stem cells was introduced by Till and McCulloch, who have shown that graded injections of bone marrow cells form 8 to 10 days later macroscopically visible nodules in the spleen of lethally irradiated mice (3). Each of these nodules containing differentiated and undifferentiated cells constitutes a colony originating from a single cell termed colony-forming-unit-spleen (CFU-S) which is capable of giving rise to myeloid, erythroid and megakaryocytic cells (4). Clonality of these colonies was initially suggested by linear relationship between cells injected and nodules formed, and was further supported by using radiation-induced chromosome markers. Injection of cells with these markers into irradiated mice results in spleen colonies showing karyotypic markers in a majority of metaphases and in both peroxydase-positive granulocytic and 55 Felabelled erythroid cells (5). These results clearly indicated the ability of a single progenitor cell for proliferation and differentiation in the three major hemopoietic cell lineages. Demonstration of self-renewal capacity of these cells was achieved by determining the number of new CFU-S generated in a second irradiated recipient animal after injection of cells coming from colonies obtained by taking out spleen of a first irradiated mouse (7). Establishing the properties of the stem cells the transplantation methodology leads to abandon earlier concepts proposing that they possess a capacity for asymmetrical division which would result in a cell becoming differentiated and in another remaining a stem cell. Similar transplantation model with stem cell deficient W/WW mice allows to demonstrate that there is a stem cell CFU-LH, for both hemopoietic and lymphoid lines. Spleen, marrow and thymus contained cells bearing chromosome markers after repopulating these animals with marrow treated for presenting irradiation-induced chromosomal markers (8). Strong evidences including success of marrow engraftment, presence of a disease-specific marker chromosome, the Philadelphia chromosome in different cell lines, studies of cell linkage with G 6 PD isoenzyme on the X chromosome, suggest the existence of a pluripotential stem cell in human (9).

While a great debate is over with animal studies demonstrating that there is a pluripotent stem cell for hematopoiesis, a steadily mounting body of data suggests that the concept of stem cell may be more complex than initially envisaged. Between pluripotent stem cell and the identifiable differentiating precursor cells, there may be an intermediate but morphologically unrecognized cell population.

COMPARTMENT OF COMMITTED PRECURSOR CELLS

First evidence concerning this intermediate class of stem cell comes from studies on stem cell kinetics with humoral factor erythropoietin. After the introduction of the spleen colony assay differences between the two assay methods were noticed and other observations led to distinguish eythropoietin — responsive cells (ERC) from CFU-S (10). Erythropoietin transforms these target cells which behave as a more differential progeny of CFU-S, to proerythroblasts. ERC are considered to be available for differentiation in the erythron and represent the first experimentally demonstrated class of committed stem cells. Further generalization of the idea of compartment of different committed stem cells comes from the development of *in vitro* clonal culture systems.

With these systems but depending on the level of erythropoietin required and the culture period, two classes of separate erythroid stem cells were identified : a cell which appeared to be analogous to the ERC, termed *burst - forming unit - erythroid* BFU-E and a more differentiated stem cell close to the proerythroblast called *colony - forming unit - erythroid* CFU-E. If different observations suggest that CFU-E is a mature descendant of the BFU-E, there is also cumulative evidences for a more primitive erythroid stem cell than the BFU-E (11).

In contrast to erythropoiesis there was no practical method for reversibly suppressing the granulocyte production, but *in vitro* culture systems have provided the analytic tool for the study of granulopoiesis. In 1965, Pluznik and Sachs, and Bradley and Metcalf independently described methods which produced maintenance of growth, yielding granulocytic colonies from explanted marrow cells (12). Cells that give rise to these colonies were termed *colony-forming unit in culture* (CFU-C). There was evidence indicating that they represented different cell populations. With the culture systems, the cells are suspended in a semi-solid medium in which microscopically visible colonies develop but they need for growth the presence of substances named colony - stimulating factor (CSF) or activity (CSA). CSF activity has been demonstrated by feeder layers of embryonic cells or appeared in media conditioned by leukocytes. These methods led to the identification of various progenitors such as the granulocyte and monocyte committed stem cells (CFU-GM) or the megakaryocytic committed stem cells... They paved the way for analytical studies and purification of the different CSF, that gave the opportunity for beginning to understand the nature of the controlling factors involved in the regulation of hematopoiesis. *In vitro* culture systems considerably enriched the concept of stem cell showing the existence of a class of intermediate progenitors without morphological descriptiveness except through their descendants (13).

GROWTH FACTORS

Stimulatory effect exerted by erythropoietin on red cell production was first suggested in 1906 by Carnot and Deflandre who hypothesized that a humoral factor was generated after arterial hypoxia. It took almost 50 years before experimental data became available to confirm this hypothesis and evidence was provided for the existence of erythropoietin (14). The erythropoietic hormone was further characterized biochemically, but the demonstration of kidney as its site of production was initially established by Jacobson and his associates who showed that bilateral nephrectomy inhibited response of rats to various forms of hypoxia, while bilateral ureteral ligation which produced an equivalent degree of uremia resulted in only a slight reduction of the increase of plasma erythropoietic activity (15).

At the end of the 1970s, conditioned media from mitogen-stimulated mononuclear cells were used for *in vitro* clonal culture systems to stimulate the proliferation of BFU-E and granulocyte, macrophage or eosinophil colonies. It was originally not clear whether one single factor or multiple molecules were involved in the different colony types (16).

It was proposed this burst-promoting activity (BPA) activates the proliferation of pluripotent stem cell and that the subsequent differentiation corresponded to an increase in pathway specific receptors. IL-3 (Interleukin 3) which seems to be identical to factors defined by their biological activities such as BPA or multi-colony stimulating factor (Multi - CSF), is a factor characterized by polymorphic actions. Stimulating CFU-S, committed progenitors and granulocyte colonies, IL-3 requires the addition of erythropoietin to obtain maximal production of eythroid colonies. The mouse IL-3 was been cloned in 1984. Cloned two years later the human IL-3 gene is located near the GM-CSF gene. The same approach has been used to identify other growth factors and to define their targets (17). If it is possible to propose a classification of these factors depending on their targets, multipotent cells or committed progenitors, many of them act simultaneously at different levels of hematopoiesis. Using molecular biology techniques the receptors of these factors have been identified. Two of them with tyrosine kinase activity are encoded by the proto-oncogenes c-fms and c-kit.

MICRO ENVIRONMENT

The influence of environment termed hematopoietic-inductive microenvironment (HIM) was suggested in 1967 by studying the distribution of colonies in the spleen after transplantation. With the development of long-term marrow cultures and biochemistry of high molecular weight molecules the understanding of this concept evolved. Microenvironment appears today consist of cells of the stroma and extra cellular matrix which play a role in the adherence of hematopoietic cells to the stroma and secretion of growth factors. Two loci identified thirty years ago (W, SI) and known to be responsible in the homozygotes of a severe anaemia with die in utero, have been recently characterized. The first one corresponds to the proto-oncogene c-kit and the second to the ligand of c-kit produced by stroma cells.

IMMUNOLOGY

The technology of monoclonal antibodies initially developed by Kohler and Milstein and cell separation methods open the way to search for direct strategy to purify different stem cells (19). If they allow to revise the scheme of differentiation based on morphology, they do not lead to identification of antigenes specific of a particular primitive progenitor. Nevertheless with the use of a variety of phenotypic markers characterization of mouse hematopoietic stem cells have been achieved recently and it has been demonstrated that thirty of these cells are sufficient to save 50 percent of lethally irradiated mice with reconstitution of all blood cell types in the survivors (20).

CONCLUSION

Two main questions have dominated the recent history of hematopoiesis, the nature of the stem cells and the mechanism of differentiation. Despite several advances fundamental questions remain concerning the regulation of the differentiation pathways and the control of self-renewal of stem cells. Three theories have been proposed to resolve these problems. One of them favors the idea that stem cell are organized for use on the basis of their generation age. The second points out the role of the micro environment and the third claims for a stochastic process maintaining an equilibrium between self-renewal and commitment.

REFERENCES

1. F.R. Sabin, F.R. Miller, K.C. Smithburn, R.M. Thomas, L.E. Hummel, " Changes in the bone marrow and the blood cells of developing rabbits ", *J. Exp. Med.*, 64 (1936), 97-120. S. Abramson, R.G. Miller, R.A. Phillips, " The identification in adult bone marrow of pluripotent and restricted stem cells of the myeloid and lymphoid system ", *J. Exp. Med.*, 145 (1977), 1567-1579. W. Vainchenker, " Hématopoïèse ", in B. Dreyfus (eds), *L'Hématologie*, Paris, Flammarion, 1992, 3-41.

2. C.E. Ford, J.L. Hamerton, D.W.H. Barnes, J.F. Loutit, " Cytological identification of radiation-chimarras ", *Nature*, 177 (1956), 452-454 ; P.J. Quesenbery, " The concept of hemopoietic stem cell ", in W. Williams, E. Beutier, P.J. Erslev, M.A. Lichtman (eds), *Haematology*, New York, Mc Graw-Hill Book Company, 1983, 129-143.

3. J.E. Till, E.A. McCulloch, " A direct measurement of the radiation sensitivity of normal mouse bone marrow cells ", *Radiat. Res.*, 14 (1961), 213-222 ; E.A. McCulloch, " Control of Hematopoiesis at the cellular level ", in A. Gordon (eds), *Regulation of Hematopoiesis, 1, Red Cell production*, New York, Meredith Corporation, 1970, 133-160.

4. J.P. Lewis, F.E. Trobaugh Jr., " Haematopoietic stem cells ", *Nature*, 204 (1956), 589-590. A.J. Becker, E.A. McCulloch, J.E. Till, " Cytological demonstration of the clonal nature of spleen colonies derived from transplanted mouse marrow cells ", *Nature*, 197 (1963), 452-454.

5. A.M. Wu, J.E. Till, L. Siminovitch, E.A. McCulloch, " A cytological study of the capacity for differentiation of normal hemopoietic colony-forming cells ", *J. Cell Physiol.*, 69 (1967), 177-184.

6. V. Juraskova, L. Tkadlecek, " Character of primary and secondary colonies of haematopoiesis in the spleen of irradiated mice ", *Nature*, 206 (1965), 951-952.

8. J.L. Curry, J.J. Trentin, " Hemopoietic spleen colony studies. I. Growth and differentiation ", *Dev. Biol.*, 15 (1967), 395-413 ; J.L. Curry, J.J. Trentin, U. Cheng, " Hemopoietic spleen colony studies. III. Hemopoietic nature of spleen colonies induced by lymph node or thymus cells, with or without phytohemagglutinin ", *J. Immunol.*, 99 (1967), 907-916.

9. J. Whang, E. Frei, J.H. Tjio, P. Carbone, G. Brecher, " The distribution of the Philadelphia chromosome in patients with chronic myelocytic leukaemia ", *Blood*, 22 (1963), 664-673 ; P.J. Fialkow, R.J. Jacobson, T. Papayannopoulou, " Chronic myelocytic leukaemia : Clonal origin in stem cell common to the granulocyte, erythrocyte, platelet and monocyte/macrophage ", *Am. J. Med.*, 63 (1977), 125-130.

10. E. Filmanowicz, C.W. Gurney, " Studies on erythropoiesis XVI, response to a single dose of erythropoietin in polycythemic mice ", *J. Lab. Clin. Med.*, 57 (1961), 65-72 ; G. Hodgson, " Mechanism of Action of Erythropoietin ", in A. Gordon (eds), *Regulation of Hematopoiesis, 1, Red Cell production*, New York, Meredith Corporation, 1970, 459-469.

11. D.L. Mc Leod, M.M. Shreeve, A.A. Axelrad, " Improved plasma culture system for production of erythrocytic colonies in vitro : Quantitative

assay method for CFU-E ", *Blood*, 44 (1974), 517-534 ; R.F. Humphries, A.C. Eaves, C.J. Eaves, " Characterization of a primitive erythropoietic progenitor found in mouse marrow before and after several weeks in culture ", *Blood*, 53 (1979), 746-763.

12. D.H. Pluznik, L. Sachs, " The cloning of normal mast cells in tissue culture ", *J. Cell Comp. Physiol.*, 66 (1965), 319-324 ; T.R. Bradley, D. Metcalf, " The growth of mouse bone marrow cells in vitro ", *Aust. J. Exp. Biol. Med.*, 44 (1966), 287-300.

13. L. Sachs, " *In vitro* control of growth and development of hematopoietic cell clones ", in A. Gordon (eds), *Regulation of Hematopoiesis, 1, Red Cell production*, New York, Meredith Corporation, 1970, 217-233 ; D. Metcalf, J. Parker, H.M. Chester, P.W. Kincade, " Formation of eosinophilic-like granulocytic colonies by mouse bone marrow cells *in vitro* ", *J. Cell Physio.*, 84 (1974), 275-290.

14. P. Carnot and C. Deflandre, " Sur l'activité hématopoiétique des sérums en cours de régénération du sang ", *CR Acad. Sci.*, 3 (1906), 384-387 ; A.J. Erslev, " Humoral regulation of red cell production ", *Blood*, 8 (1953), 349-357.

15. T. Miyake, C.K.H. Kung, E. Goldwasser, " Purification of human erythropoietin ", *J. Biol. Chem.*, 252 (1977), 5558-5564. L.O. Jacobson, E. Goldwasser, W. Fried, L. Plzak, " Role of the kidney in erythropoiesis ", *Nature*, 179 (1957), 633-634 ; A.S. Gordon, E.D. Zanjani, " Some aspects of erythropoietin physiology ", in A. Gordon (eds), *Regulation of Hematopoiesis, 1, Red Cell production*, New York, Meredith Corporation, 1970, 413-457.

16. G.R. Johnson, D. Metcalf, " Nature of cells forming erythroid colonies in agar after stimulation by spleen conditioned medium ", *J. Cell Physiol.*, 94 (1978), 243-252.

17. J. Suda, T. Suda, K. Kubota, J.N. Ihle, M. Saito, Y. Miura, " Purified interleukin-3 and erythropoietin support the terminal differentiation of hemopoietic progenitors in serum-free culture ", *Blood*, 67 (1986), 1002-1006 ; M.C. Fung, A.J. Hapel, S. Ymer, D.R. Cohen, R.M. Johnson, H.D. Campbell, I.G. Young, " Molecular cloning of c-DNA for mouse interleukin-3 ", *Nature*, 307 (1984), 233-236.

18. E. Huang, K. Nocka, D.R. Beier, T.Y. Chu, J. Buck, H.W. Lahm, D. Wellner, P. Leder, P. Besmer, " The hematologic growth factor KL is encoded by Sl locus and is the ligand of c-kit receptor, the gene product of the W locus ", *Cell*, 63 (1990), 225-233.

19. G. Kohler, S. Milstein, " Continuous cultures of fused cells secreting antibody of undefined specificity ", *Nature*, 256 (1975), 495-497.

20. G.J. Spangrude, S. Heifeld, I.L. Weissaman, " Purification and characteristics of mouse hematopoietic stem cells ", *Science*, 241 (1988), 58-62.

HISTORY OF ALLOMETRY

Jean GAYON

The word " allometry " was coined by Julian Huxley and Georges Teissier in 1936 (Huxley & Teissier 1936a, 1936b). It referred to a law of relative growth which had been discovered earlier by several authors. This paper sketches out the story of this law. Julian Huxley is often said to have discovered the quantitative law of constant differential growth in 1924, but Teissier recalled in 1926 that this mathematical formula had been already applied to mammals nearly three decades before by Dubois and Lapicque. This paper examines : (1) the line of research that led to Huxley's discovery ; (2) Teissier's contribution and historical reconstruction of the story of the discovery ; (3) the 1936 joint papers by Teissier and Huxley.

PÉZARD, CHAMPY, HUXLEY

Since the 1900s, several biologists had observed that, in many animals, secondary sexual characters grow relatively larger and larger as the animal grows. In 1918, Albert Pézard (1875-1927) provided the first experimental and quantitative treatment of this problem. Pézard worked on Gallinaceae. He showed that the physiological determinism of secondary sexual characters in cocks is not simple : castration stops the development of some of them (especially the crest), but not all : spurs do not depend on testicles (Pézard, 1911). Spurs develop spontaneously ; this is shown by the observation that removal of ovaries in female causes the development of spurs. In 1918, Pézard gave a quantitative description of these phenomena, and illustrated them by remarkable graphs. He plotted the length of crest and spurs against the body size, measured by the cubic root of weight ($\sqrt[3]{P}$). The empirical curves showed an obvious " discordance " between the curves of body size and crest size, whereas the growth of the spurs followed approximately the general development of the organism. It is on this occasion that Pézard coined the expressions " isogonic growth " (for organs the growth of which follows the general development of the organism) and " heterogonic growth " (discordant growth). " Heterogonic growth " remained the commonest expression for this kind of phenomena until the introduction of " allometry " in 1935, especially in the English literature. Pézard's remarkable experimental studies influenced a lot of people in various areas : physiology of sex, embryology, endocrinology, biometry. The most important point in his study was that the relevant variable for relative growth was not time, but body size. There was however a major limitation in Pézard's inquiry. He did not offer any algebraic form of the law of heterogonic growth of the crest.

Champy was a physician, who had an influential role in France in the inter war years. In 1924, in a book entitled on sexuality and hormones, he coined the expression " Dysharmonic growth " for " an extremely general phenomenon " which he claimed to have established : the phenomenon of the continuous increase of the relative size of secondary sexual characters in function of the body size (Champy, 1924, 1929). The 1924 book provided impressive illustrations of this phenomenon, especially in Insects. Champy believed that differential growth in mature animals was a phenomenon almost always restricted to secondary sexual characters. He explained this phenomenon by a sexual hormone acting specifically on certain parts of the body. He thought that a given hormone always causes a similar acceleration of cell multiplication in some sensitive areas, and thus induces a growth process which is adequately described by a parabolic curve. He thus proposed that " dysharmonic growth " (Pézard's " heterogony ") followed a parabolic law of the form : $V = at^2$ (where 'V' is a measure of the secondary sexual character, 't' is the body size, and 'a' a constant). In this formula, the relative growth of an organ is a function of body size.

Champy's book was written and published just before the famous paper in which Huxley (1924) proposed his own law of differential growth, which is in fact no more than a generalization of Champy's law, since it is a power law (Huxley's formula is : $y = bx^k$). Actually, Huxley's paper of 1924 quotes both Pézard (1918) and Champy (1924). Huxley constantly referred to Champy as a major contributor to the resolution of the problem of relative growth.

Julian Huxley (1887-1975) is primarily known for his paradigmatic book *Evolution - The Modern Synthesis* (1942). Nevertheless his major scientific accomplishment was certainly his contribution to the study of relative growth. Huxley's first paper on relative growth appeared in 1924. It tried to answer to a question raised by Thomas H. Morgan, a year earlier, on the abdominal width of female fiddler crabs (Uca pugnax). Morgan was puzzled by the very large abdomen of some of these animals, and wondered whether this character resulted from their genetic make-up or from the law of growth. Working on Morgan's data, Huxley argued in favor of the second hypothesis, and, on this occasion, used for the first time Pézard's terminology of " heterogonic " and " isogonic " growth. Although this paper does not provide the law of heterogony which made him famous a few months later, it already gave a very simple method for detecting heterogonic growth : " The best method of detecting and analyzing heterogonic growth-rate is by plotting the percentage size of the part in question against the absolute size of some dimension of the whole body " (Huxley, 1924a ; 475).

Nevertheless, this paper did not say anything about the law of heterogonic growth. This was precisely the object of a second paper, published a few months later in *Nature*. In this paper, Huxley stated a law of heterogonic growth for the chelae of fiddler-crabs. This law is a power law of the form : $y = bx^k$. Or, using logarithms : $\log y = k \log x + \log b$ (y, magnitude of the differentially-growing organ ; x , size of the body ; k, constant differential growth-ratio ; b, constant [origin index]). The essential theoretical feature of this formula is the following : what is constant (k) is not a ratio of two sizes but a ratio between two growth-rates. Although form is not constant, change of form can be constant. This is the significance of Huxley's law.

Huxley believed he had discovered a major empirical law in biology. As he himself said, and as Teissier insisted later, the physical basis of the law was mysterious. But the generality of its application, for many characters in animals and plants, made it a very serious generalization in biological sciences. In practice, because the power law could be expressed on a double logarithmic scale, it provided a remarkably easy method for detecting and proving the existence of heterogonic growth. In fact, soon after Huxley discovered his law of differential growth, it was widely verified by himself and others. A genuine industry of differential growth began to flourish at the end of the 1920s. It generally consisted in exhibiting a logarithmic graph of relative growth of some character or other. This industry finally led Huxley to publish a synthetic book on relative growth in 1932. Huxley himself provided himself a number of verifications of the law, but the subject rapidly became an international enterprise. Instead of analyzing Huxley's synthesis, I will now consider Teissier's approach, which is similar up to a certain point, but which also reveals other interesting features of the history of conceptions about differential growth.

TEISSIER

Georges Teissier (1900-1972) was fifteen years younger than Huxley. Mathematically trained, he was both interested in systematics and biometry. When Huxley discovered his law of heterogony in 1924, Teissier had not yet written anything on biometry. In his doctoral dissertation of 1931, devoted to quantitative approaches of the growth of insects, he said that he had came to the power law independently. Teissier's first paper on relative growth appeared in 1926. This paper bore upon the size of ommatidia (the units of an insect's compound eye) as a function of body size in various insects. It showed that " in a definite species… the bigger the insect is, the bigger the facets of the eye are ". Teissier refereed to Champy's " dysharmonic growth " (Teissier, 1926). This paper also gave an algebraic formula, with the form of a power law. However it did not refer to Huxley, but to a French scientist named Louis Lapicque, who had compared the size of the eye with body size in vertebrates (Lapicque & Grioud, 1923). In his following papers on differential growth, Teissier continued to refer to Lapicque. But he also began to refer to Huxley and use a formula of differential growth which was formally identical with Huxley's. Still he never said that Huxley had discovered it. Finally, in his 1931 dissertation, Teissier devoted a full paragraph to the history of relative growth. There he acknowledged the important role of Huxley, but denied that Huxley had discovered the method of describing differential growth with the aid of a power law and logarithmic coordinates. He said that this method of description of relative growth had been discovered in 1897 and 1898 by Dubois (Dutch) and Lapicque (French), and successfully applied to the study of the variation of characters such as brain size or the area of the retina, as a function of body size in vertebrates (Teissier 1931 : 88-93). I have checked some of Teissier's sources. He was right : since 1897, there was a continuous tradition of using power laws and logarithmic coordinates in order to describe phenomena of typical relative growth. From 1897 to 1923, this method was used in an impressive series of papers (Dubois 1897 ; Lapicque 1898, 1907, 1908, 1910 ; Lapicque & Grioud 1923). Did Teissier deliberately quote Lapicque instead of Huxley in his first paper, in order to avoid recognizing Huxley's priority ? This is possible. But even if this were true, the net result is that the real priority does not belong to Huxley. Did Huxley know about Dubois and Lapicque when he suddenly proposed his law of constant differential growth in 1924 ? I have no special evidence on this point. Huxley, in his 1932 synthetic book on relative growth, quotes various papers by Dubois or Lapicque.

HUXLEY'S AND TEISSIER'S JOINT PAPER (1936)

In 1935, Huxley and Teissier decided to agree on a common terminology for relative growth. During a few months, they exchanged letters, and negotiated various compromises regarding the designation, vocabulary and symbolic notation of the law of relative growth. Two joint papers were published in 1936 in French (*Comptes rendus de la Société de biologie*) and in English (*Nature*). The two authors decided to give up the two terms they respectively used for the designation of the phenomenon : " allometry " replaced Huxley's " heterogony " and Teissier's " dysharmony " ; " isometry " replaced " isogony " and " harmony ". They also agreed on a common symbolic formulation of the law : $y = bx^a$.

The comparison of the French and the English version show that most differences are unessential. There is however one major difference. It concerns the constant 'b'. For Huxley, this constant had no biological significance whatsoever. 'b' was no more than the value of x when y = 1. This constant was therefore arbitrary, and depended only of the choice of the measuring-unit. Teissier did not agree with this view. For him, the constant 'b' could be given a biological significance if one paid attention to the statistical nature of the data. For this reason, he introduced in the French version the following sentence : " From a statistical point of view, [b] represents the mean value of the ratio y/x for all the observed individuals " (Huxley & Teissier, 1936a : 936. For a more elaborate justification, see also Teissier 1934 : 301). Huxley did not introduce this sentence in the English version.

The correspondence shows that this disagreement had something to do with the distinction between intraspecific allometry (or allometry of growth) and inter specific allometry (or allometry of size). Teissier thought these two phenomena were biologically distinct, and required a different symbolic notation ; Huxley agreed to mention the distinction, but did not accept different symbols.

In fact this was a very old problem. In 1898, when Lapicque had first stated the necessity of using a power law in order to describe the relation between brain size and body size in mammals, he had already indicated an interesting fact : for a given set of species, the differential growth ratio is not the same within a species as between species.

CONCLUSION

In the domain of allometry the quarrel of priority is hopeless. Champy claimed to have first discovered the law of dysharmonic growth. But he did not offer the correct formula. Huxley claimed to have discovered that the law of relative growth was a power law. But Teissier, who himself came to the same law, found it in Dubois and Lapicque thirty years earlier. Had these people been 'precursors' ? Yes, in the sense that they had given the right formula. No because it is doubtful whether Dubois and Lapicque really intended to provide a general law of *growth*. The history of allometry involved indeed many cumulative efforts of many people. The problem involved quite serious formal and empirical constraints, which had to be thoroughly understood and applied to a large series of biological objects. This is why the story seems to repeat itself, although it did not. Several people 'discovered' the law of allometry, with various methods and theoretical objectives. The discovery belongs to all of them. This being said, Julian Huxley, even if he did not play the role of absolute beginner, certainly had the major role.

ACKNOWLEDGEMENTS

I thank Richard M. Burian for his useful comments and linguistic corrections.

BIBLIOGRAPHY

1) Unpublished sources

Correspondence between Huxley and Teissier, 1935-36, Transcription : Michel Veuille, Location : Laboratoire de génétique des populations, Université Paris 6 Pierre et Marie Curie, 7 quai Saint-Bernard, 75252 Paris.

2) Published material

C. Champy, 1922a, *L'action de l'extrait thyroïdien sur la multiplication cellulaire. Caractère électif de cette action. Archives de morphologie générale et expérimentale 1. Separate instalment* ; Paris, Doin, 1922, 58 p.

C. Champy, 1922b, *Exposé des titres et travaux scientifiques de M. Christian Champy 1906-1922*, Paris, Doin, 84 p.

C. Champy, *Sexualité et hormones*, Paris, Doin, 1924.

C. Champy, " La croissance dysharmonique des caractères sexuels accessoires ", *Archives des sciences naturelles - Zoologie,* 12 (1929), 193-244.

C. Champy, *Résumé des principales recherches 1906-1952*, Roneoed document (Muséum national d'histoire naturelle de Paris), 1952, 10 p.

E. Dubois, " Sur le rapport du poids de l'encéphale avec la grandeur du corps chez les Mammifères ", *Bulletin de la Société d'Anthropologie de Paris,* 8 [4] (1897), 337.

J.S. Huxley, " The Variation in the Width of the Abdomen in Immature Fiddler Crabs considered in relation to its Relative Growth-Rate ", *The American naturalist,* 58 (1924a), 468-475.

J.S. Huxley, " Constant Differential Growth-Ratios and their Significance ", *Nature,* 114 (1924b), 895-896.

J.S. Huxley, *Problems of Relative Growth,* New York, Lincoln Mac Veagh - The Dial Press, 1932, 276 p.

J.S. Huxley and G. Teissier, " Terminology of Relative Growth ", *Nature,* 137 (1936a), 780-781.

J.S. Huxley and G. Teissier, " Terminologie et notation dans la description de la croissance relative ", *Comptes Rendus des Séances de la Société de Biologie,* 121 (1936b), 934-937.

E. Lapicque, "Sur la relation du poids de l'encéphale au poids du corps ", *Comptes rendus des séances de la Société de Biologie,* 10th series, 5 (1898), 62-63.

E. Lapicque, " Tableau général des poids somatiques et encéphaliques dans les espèces animales ", *Bulletin de la Société d'Anthropologie de Paris,* 8 [5] (1907), 248.

E. Lapicque, " Le poids encéphalique en fonction du poids corporel chez les individus d'une même espèce ", *Bulletin de la Société d'Anthropologie de Paris,* 8 [5] (1908), 313.

E. Lapicque, " Relation du poids encéphalique à la grandeur rétinienne dans quelques ordres de mammifères ", *Comptes rendus hebdomadaires des séances de l'Académie des sciences,* 111 (1910), 1393-1396.

E. Lapicque et Grioud, " En fonction de la taille de l'animal le nombre des neurones sensitifs varie moins vite que celui des neurones moteurs ", *Comptes Rendus des Séances de la Société de Biologie,* 89 (1923).

T.H. Morgan, " Further evidence on Variation in the Width of Abdomen in Immature Fiddler-Crabs ", *The American Naturalist,* 57 (1923), 274.

A. Pézard, " Sur la détermination des caractères sexuels secondaires chez les Gallinacés ", *Comptes rendus hebdomadaires des séances de l'Académie des sciences,* 153 (1912), 1027-1029.

A. Pézard, " Sur la détermination des caractères sexuels secondaires chez les Gallinacés : greffe de testicule et castration post-pubérale ", *Comptes rendus hebdomadaires des séances de l'Académie des sciences,* 154 (1912), 1183-1186.

A. Pézard, " La loi numérique de la régression des irgabes érectiles, consécutive à la castration postpubérale, chez les Gallinacés ", *Comptes rendus hebdomadaires des séances de l'Académie des sciences,* 164 (1917), 734-736.

A. Pézard, " Le conditionnement physiologique des caractères sexuels secondaires chez les oiseaux ", *Bulletin biologique de la France et de la Belgique,* 52 (1918), 1-176.

G. Teissier, " Sur la biométrie de l'œil composé des insectes ", *Bulletin de la Société zoologique de France,* 51 (1926), 501-505.

G. Teissier, " Sur les dysharmonies de croissance chez le Insectes ", *Comptes rendus des séances de la Société de Biologie,* 99 (1928), 297-298.

G. Teissier, " Sur quelques dysharmonies de croissance des crustacés brachyures ", *Comptes rendus des séances de la Société de Biologie,* 99 (1928), 1934-1935.

G. Teissier, " Dysharmonies biochimiques dans la croissance larvaire de Tenebrio molitor L ", *Comptes rendus des séances de la Société de Biologie,* 100 (1928), 1171-1173.

G. Teissier, " La croissance embryonnaire de Chrysaora hysocella (L.) ", *Archives de zoologie expérimentale et générale,* 69 (1929), 137-178.

G. Teissier, " Recherches morphologiques et physiologiques sur la croissance des insectes ", *Travaux de la Station Zoologique de Roscoff,* Fasc. 9 (1931), 29-238.

G. Teissier, " Description quantitative de quelques croissances complexes ", *Annales de physiologie et de physicochimie biologique,* 10 (1934a), 359-376.

G. Teissier, *Dysharmonies et discontinuités dans la croissance,* Paris, Hermann, 1934b, 39 p.

G. Teissier, " Les procédés d'étude de la croissance relative ", *Bulletin de la Société zoologique de France,* 60 (1935), 292-307.

GEOFFROY SAINT-HILAIRE REVISITÉ

Hervé LE GUYADER

De récents articles de biologie du développement, publiés dans des journaux scientifiques prestigieux comme *Nature* (1) ou *Current Biology* (2), font explicitement référence aux idées hétérodoxes qu'Etienne Geoffroy Saint-Hilaire avait émises au début du XIXᵉ siècle. L'article " Considérations générales sur la vertèbre " (3) de 1822 est le plus souvent cité. Geoffroy Saint-Hilaire y présente une identité de plan entre les arthropodes (animaux hyponeuriens, dont le système nerveux est ventral) et les vertébrés (animaux épineuriens, dont le système nerveux est dorsal). Il va même jusqu'à postuler une inversion dorso-ventrale des vertébrés par rapport aux arthropodes et affirmer que ces derniers sont des " dermo-vertébrés ", ce qui déclencha immédiatement une série de remarques acerbes de la part de Georges Cuvier.

Cet article est abondamment cité à propos des dernières découvertes en génétique moléculaire du développement concernant les organisations antéro-postérieure et dorso-ventrale des animaux triblastiques. En effet, ceux-ci ont leurs polarités essentielles générées par les mêmes systèmes génétiques, indépendamment de leurs plans d'organisation. Les gènes *Hox* ont pour rôle la structuration antéro-postérieure de l'ensemble des animaux triblastiques ; la polarité dorso-ventrale quant à elle est déterminée chez la drosophile par le système *sog/dpp*. Chez le crapaud xénope, c'est le système homologue *BMP-4/chordin* qui joue ce rôle ; le point-clé est qu'on trouve des expressions inversées des gènes homologues chez ces deux animaux, c'est-à-dire que le gène qui détermine la face ventrale de l'un s'exprime au niveau de la face dorsale de l'autre, et *vice-versa*.

Peut-on réellement dire comme certains que la génétique du développement retrouve les idées émises un siècle et demi plus tôt par Etienne Geoffroy Saint-Hilaire ? Afin de répondre à cette interrogation, il convient de revoir succinctement les principales conceptions du grand zoologiste. L'analyse ne doit pas porter uniquement sur les " Considérations générales sur la vertèbre " de 1822, mais doit également considérer les " Mémoires sur les insectes " de 1820 —

surtout le premier des trois (4) — souvent passés sous silence. En effet, dans cet article, on trouve une bien étrange phrase qu'il convient d'expliquer : " Tout animal habite en dedans ou en dehors de sa colonne vertébrale ".

Pour arriver à une telle conclusion, Geoffroy Saint-Hilaire avait réalisé une comparaison entre vertébrés et arthropodes — pour lui le monde animal semblait à ce moment s'arrêter là — et proposé une identité d'origine et de fonction entre la cuticule des insectes et la colonne vertébrale des vertébrés (4) : " on trouve chez les insectes, à la fois contenus dans le même tube, non seulement leur moelle épinière, mais tous les organes abdominaux ". Par conséquent : " De ces faits, il y a à conclure que les insectes sont des animaux vertébrés ". Pour arriver à cette conclusion, Geoffroy Saint-Hilaire n'a fait que retourner un homard et proposer une analogie floue entre arthropodes et vertébrés : " Je plaçai l'animal non plus comme il est posé relativement au sol, mais comme il me convenait de le voir pour le comparer aux animaux des premiers rangs. [...] Quelle fut ma surprise [...] en apercevant une ordonnance qui plaçait sous mes yeux tous les systèmes organiques de ce homard dans l'ordre où ils sont rangés chez les animaux mammifères ? ".

Il faut bien insister sur le fait que la comparaison porte sur des animaux adultes, donc sur des plans d'organisation. C'est un point crucial, car il montre à quel point on peut arriver à une incompréhension de Geoffroy Saint-Hilaire, quand on présente les derniers acquis de la génétique du développement comme étant une redécouverte de la pensée du grand ancien.

On peut arriver à de plus grandes erreurs encore si on oublie que Geoffroy Saint-Hilaire, né en 1772 et mort en 1844, ne pouvait avoir connu le darwinisme. Ainsi, malgré de louables efforts, il ne se sera jamais complètement débarrassé du concept de l'Echelle des Etres. C'est ainsi qu'il pensait que, après fécondation, chaque embryon initiait le même développement, l'un s'arrêtant tôt, l'autre allant plus loin dans ce processus, donnant naissance à un mollusque, un insecte ou un vertébré. Cette idée de base le mena à proposer que chaque animal répondait à un même pattern de base. Il y a une seule et même " carrière " pour chaque embryon, mais cette carrière est donnée une fois pour toute. Elle relève en fait d'une transcendance.

Maintenant on peut comprendre la réponse que Geoffroy Saint-Hilaire fit à Georges Cuvier lors de la célèbre discussion de 1830 à l'Académie des Sciences (5). En effet, Cuvier fit la remarque que bon nombre d'animaux n'ont pas d'hyoïde, et que par conséquent on ne peut parler de cet organe de manière générale. La réponse de Geoffroy Saint-Hilaire est entièrement contenue dans le cadre de cette réflexion : " Il faut heure, âge convenable pour que, dans un embryon quelconque, d'homme, de mammifère, d'oiseau, etc., l'hyoïde apparaisse ; auparavant il n'est pas compatible avec le degré d'organisation de cette époque. De même chez les animaux qui appartiennent à ce même degré des développements organiques, il n'y a, il ne peut y avoir d'hyoïde ; quoi de surprenant à cet égard ? " Ainsi, pour Geoffroy Saint-Hilaire, si l'hyoïde n'est pas obligatoirement là, il y est potentiellement.

Il paraît donc clair que la lecture actuelle que l'on peut avoir de Geoffroy Saint-Hilaire est souvent très approximative. Bien sûr, il a proposé l'existence d'une inversion dorso-ventrale entre arthropodes et vertébrés, mais à partir d'une analyse d'anatomie comparée qui s'est vite avérée erronée. Il n'y a ici aucun doute, c'est par une comparaison de plan d'organisation que Geoffroy Saint-Hilaire a procédé.

A l'inverse, la vision moderne s'interprète suivant un concept de prepattern, et dans un cadre darwinien. En effet, si l'on reconnaît que l'axe antéro-postérieur d'une part, l'axe dorso-ventral d'autre part, sont déterminés par des systèmes génétiques homologues chez les vertébrés et arthropodes, on peut remarquer que (i) ces systèmes sont activés très tôt au cours de l'embryogenèse, en tout cas avant ou pendant le moment où le stade phylotypique est acquis, c'est-à-dire avant ou lorsque les grandes caractéristiques du plan d'organisation apparaissent d'un point de vue phénotypique ; (ii) l'interprétation évolutive s'inscrit dans un cadre darwinien, à savoir que ces animaux ont hérité ces systèmes génétiques d'un ancêtre hypothétique commun. De plus, l'intérêt évolutif de ces systèmes est que, malgré leur homologie, ils permettent l'instauration de plans d'organisation différents. Postuler une inversion dorso-ventrale de plans d'organisation n'est conceptuellement pas équivalent à distinguer une inversion dorso-ventrale de l'expression de gènes homologues du développement.

Néanmoins, il faut reconnaître la pertinence de certaines des intuitions d'Etienne Geoffroy Saint-Hilaire. Celles-ci ont l'intérêt d'apparaître comme des programmes prometteurs de recherche, et beaucoup sont actuellement encore considérées d'un intérêt secondaire. En effet, par l'étude de prepattern d'expression, on oublie la forme ; or celle-ci est quand même essentielle pour comprendre l'adaptation des organismes à leur milieu.

RÉFÉRENCES

1. E.M. De Robertis & Y. Sasai, " A common plan for dorsoventral patterning in Bilateria ", *Nature*, 380 (1996), 37-40.
2. C.M. Jones & J.C. Smith, " Revolving vertebrates ", *Current Biology*, 5 (6) (1995), 574-576. Le titre peut également se traduire par : les vertébrés qui se retournent.
3. E. Geoffroy Saint-Hilaire, " Considérations générales sur la vertèbre ", *Mém. Mus. Hist. nat.*, 9 (1822), 89-114.
4. E. Geoffroy Saint-Hilaire, " Mémoire sur l'organisations des Insectes ", Premier mémoire sur un squelette chez les Insectes dont toutes les pièces identiques entre elles dans les divers ordres du système entomologique correspondent à chacun des os du squelette dans les classes supérieures

(lu à l'Académie des sciences, le 3 janvier 1820), *Journ. complém. du Dict. des Sc. méd.*, 5 (1819), 340.

5. E. Geoffroy Saint-Hilaire, *Principes de philosophie zoologique, discutés en mars 1830, au sein de l'Académie Royale des Sciences*, Paris, Pichon et Didier, 1830.

HOMOLOGIES OF PROCESS

Scott F. GILBERT

During the 1990s, discoveries in developmental genetics indicated that some of the most critical phenomena in the animal kingdom are underlain by the same genes and processes. The "molecular toolkit" was found to be very small, and developmental events as disparate as the formation of the nematode vulva and the fly photoreceptor used similar pathways. Just as all the proteins in the animal kingdom are derived from only twenty amino acids, it appears that a large number of tissues are formed from a relatively small number of signal transduction pathways. Structures having no functional or genealogical relationship could be formed using a homologous cassette of genes. This had widespread implications for our view of the roles of genes in evolution.

The concept of the gene has undergone remarkable changes since the Modern Synthesis was formulated in the 1940s. First, the gene of the Modern Synthesis was an abstraction. It did not even matter if the gene were made of DNA or protein, since evolution was based on the transmission of selectable alleles. The gene of the Developmental Synthesis (to use a convenient shorthand) has a definite sequence of nucleotides, arranged as exons, introns, 5' and 3' untranslated regions, promoters, enhancers, silencers, and insulators. Second, the genes of the Modern Synthesis were manifest by their differences. The genes of the Developmental Synthesis are manifest by their homologies. Thus, the gene of the Modern Synthesis was the motor of natural selection. The gene of the Developmental Synthesis provides evidence for community of type. In the Modern Synthesis, the coding regions were critical, as they determined whether the enzyme was functional or non-functional, slow or fast. In the Developmental Synthesis, the regulatory regions of the gene are critical, as they determine when and where the gene is expressed. A crucial difference between the genes of the modern synthesis and current developmental syntheses concerns the places where these genes are expressed. The genes of the Modern Synthesis were expressed in adults competing for reproductive advantage. The genes of the Developmental Synthesis are expressed during the construction of embryos. This relates directly to Waddington's division (1953) of natural selection and normative selection (in adults) and stabilizing selection (in embryos).

One of the most profound differences between the genes of the Modern Synthesis and those of the Developmental Synthesis involves their autonomy. The gene of the Modern Synthesis was an atomic individual, each gene functioning independently of other genes. In the Developmental Synthesis, gene expression is not autonomous, but is part of a regulatory pathway. The genes are linked in functional ensembles. Thus, when MyoD is absent, another gene, myf5, takes its place. Myf5 is usually suppressed by the MyoD protein. (Such a gene could not be found by mutational analysis). The genes are both agents and acted upon.

This paper takes the view that there are developmental pathways that are themselves homologous both within an organism and between phyla. This goes beyond the several developmental regulatory genes (*Pax6, finge, tinman*, the *Hox* genes) that have been found to be extremely well conserved throughout animal evolution (see Gilbert, 1997). The molecular homology of the Pax6 and other genes mixes function and structure, and neither expression pattern nor structure is a good criterion for homology. (The combination of both has been used to support a homologous relationship) (see Gilbert and Bolker, 1999). However, sequence similarity can come by convergence to a common function or by the accumulation of exons from unrelated genes. Moreover, the ability of different tissues to use different mechanisms of formation (such as the lens coming from induction, autonomous specification, or regeneration from the iris) would argue against the use of expression patterns as evidence for homology.

However, whereas the structure and expression pattern of one set of genes may not be an indication of homology, the coordinated assembly of several genes and gene products into functional cassettes does indicate homologous relationships if these casettes are shared between species or between tissues within an organism. Here we get into a new understanding of homology, the homology of process.

The concept of process homology can be traced to the observations of Howard Schneiderman. He observed homologous specification — the phenomena wherein transdifferentiating cells or the products of homeotic gene mutations have a pre-existing positional information relevant to where they reside in the imaginal disc. A piece of antenna tissue that differentiates as a leg will differentiate according to its location in the antennal disc.

Schneiderman's observations (see Postelethwait and Schneiderman, 1971) were particularly important because they helped show that homology had to be defined at a particular level. Structures with no anatomical homology — the eye and the leg — may have an underlying homology of process. Hence his term, homologous specification.

We are now becoming aware of the process homology that unites the eye, the wing, and the leg discs. It is the hedgehog/Wnt pathway. These two paracrine factors interact within the disc to specify the proximal/distal, dorsal/ventral, and anterior/posterior axes. The same molecules that specify these axes in the eye also specify them in the leg and wing discs. So we have a serial process homology. Moreover, the same pathway exists in vertebrates. Every member of the pathway in insects has a homologue in the vertebrate embryo, and the same interactions that transmit the *Drosophila* Wingless signal to the nucleus through armadillo and pangolin protein are seen in the vertebrates, wherein the Wnt signal is manifest in the entry of beta-catenin and Lef-1 into the nucleus. The genes are the same and the protein interactions are the same. Only the readout is changed from tissue to tissue and from species to species. Interestingly, the same Wnt/hedgehog interactions seen in producing the fly limbs are seen in the interactions that are involved in the morphogenesis of vertebrate limbs. If a vertebrate hedgehog protein (which is usually synthesized only in the posterior mesoderm) is expressed anteriorly, the limb develops a mirror-image duplication. This is the same phenomenon that occurs when hedgehog protein is induced to form in the anterior portion of the fly wing disc (se Ingham, 1994).

We are finding many of these homologous pathways composed of the products of homologous genes. The first pathway discovered was the RTK-RAS pathway. Here, the same " cassette " is seen to be serially homologous in different regions of the *Drosophila* embryo and is evolutionarily (specially) homologous between the *Drosophila* termini, the mammalian skin, and the nematode vulva. Each uses the same modules to accomplish different ends.

Another pathway is the chordin/BMP4 pathway. Here, chordin secreted by the organiser of amphibian embryos binds to and blocks the action of BMP-4. This prevents the ectoderm from expressing the genes that specify the cells to become epidermal. Similarly, in the ventral surface of the fly embryo, the arthropod homologue of chordin (the short-gastrulation protein ; sog) blocks the lateralizing effects of its BMP-4 homologue (the decapentaplegic protein). These molecules can even substitute for their homologues. Chordin MRNA will cause neural formation in flies ; injection of sog into *Xenopus* causes ectopic notochord and CNS). We are back to Geoffroy's lobster ! BMP/chordin specifies the ectoderm to be neural, whether it be dorsal in the frog or ventral in the fly (for review and references, see De Robertis and Sasai, 1996 ; Gilbert, 1997).

Most discussions of homology have not discussed *processes* as being homologous. This is to be expected, since evolutionary biology, and systematics has been predominantly a study of adult structures. But if homologies are to be observed in the embryo, they should be homologies of processes, not structures. It was C.H. Waddington who emphasized the importance of processes. As Waddington (1975) wrote : " As far as scientific practice is concerned, the lessons to be learned from Whitehead were not so much derived from his discussions of experiences, but rather from his replacement of " things " by processes which have an individual character which depends upon the " concrescence " into a unity of very many relations with other processes ".

In terms of embryos, Waddington envisioned the concrescences of numerous genes and their products into a stabilized pathways. These developmental pathways were real " things ", and they were subject to natural selection. Whitehead was much respected by the embryologists of the 1930s because he viewed the concept of " becoming " as superior to that of " being ". As Whitehead (1933) viewed reality, " the very essence of actual reality — that is, of the completely real — is process. Thus each actual thing is only to be understood in terms of its becoming and perishing... The process is itself the actuality... " Today, we are extending Waddington's Whiteheadian notion that developmental pathways are themselves real and selectable things by noting that like " things " — limbs, ribs, wings, nucleotide sequences, amino acid sequences — they can be homologous both within and between organisms.

Homologies of developmental processes are critical modules. Studying how they are replicated and change over time will enable us to better understand the manner by which changes in development can cause the evolution of new groups of animals.

REFERENCES

E.M. De Robertis and Y. Sasai, " A common plan for dorsoventral patterning in Bilateria "*Nature* , 380 (1996), 37-40.

S.F. Gilbert, *Developmental Biology*, (Fifth ed.), Sunderland, Sinauer Associates, 1997.

S.F. Gilbert and J.A. Bolker, Homologies of Process : " Molecular Elements of Embryonic Construction ", in E. Wagner (ed.), The Character Concept in Evolutionary Biology, New Haven, Yale University Press, in press.

S.F. Gilbert, J. Opitz and R.A. Raff, " Resynthesizing evolutionary and developmental biology ", *Dev. Biol.*, 173 (1996), 357-372.

P.W. Ingham, " Hedgehog shows the way ", *Curr. Biol.*, 4 (1994), 345-350.

J.H. Postlethwait and H.A. Schneiderman, " Pattern formation and determination in the antenna of the homeotic mutant *Antennapedia* of *Drosophila melanogaster* ", *Dev. Biol.*, 25 (1971), 606-640.

C.H. Waddington, " Epigenetics and Evolution ", *Soc. Exper. Biol. Symp.*, 7 (1953), 186-199.

C.H. Waddington, *The Evolution of an Evolutionist*, Ithaca, New York, Cornell University Press, 1975, 3.

A.N. Whitehead, *Adventures of Ideas*, NY., Macmillan, 1933, 274-276.

LE BRICOLAGE DANS L'ÉVOLUTION

Avital WOHLMAN

Dans le livre qu'il vient de publier en mars 1997, François Jacob explique le lien intime qui rapproche l'homme de science de l'artiste : " malgré des moyens d'expression très différents entre le poète et le scientifique, l'imagination opère de la même façon. C'est souvent l'idée d'une métaphore nouvelle qui guide le scientifique "[1]. En effet il y a vingt ans cette année, paraissait l'article " Evolution and tinkering "[2], dans lequel F. Jacob proposait une métaphore nouvelle pour rendre compte du processus de l'évolution. Le propos de la présente réflexion est de rendre compte de cette métaphore, de sa force inspiratrice pour de nombreux chercheurs et enfin du juste milieu qu'elle nous propose.

1. LA MÉTAPHORE

Jacob a suggéré que le processus de l'évolution se réalise grâce à une activité combinatoire qui ressemblerait à celle du bricolage. A la différence du bricoleur humain dont le propos est de fabriquer ce dont il a besoin, l'activité combinatoire n'a aucun but. A l'instar de celle du bricoleur, l'activité combinatoire comporte deux aspects : le recyclage et l'improvisation. Improvisation connote ingéniosité et donc la possibilité de faire du neuf : la réorganisation des facteurs déjà existants permettrait un fonctionnement nouveau. Le recyclage souligne le conservatisme des motifs et la parcimonie des éléments qui demeurent, quelle que soit la nouveauté. D'une certaine manière la sélection naturelle passe au deuxième plan. L'accent est mis sur la multifonctionnalité des gènes et de leurs produits, au niveau de la cellule déjà mais surtout au niveau des organismes multicellulaires. Des unicellulaires aux multicellulaires, la différence principale est la capacité même de changer. Aussi, la combinatoire impose de nouvelles contraintes : plus le bricolage avance, plus se multi-

---

1. F. Jacob, *La souris, la mouche et l'homme,* Paris, Editions Odile Jacob, 1997, 223.
2. F. Jacob, " Evolution and Tinkering ", *Science,* 196 (1977), 1161-1167.

plient les contraintes, au point que le tout semble doté d'une apparente finalité. La métaphore du bricolage décrit bien selon Jacob les deux phénomènes majeurs de l'évolution que sont l'analogie et l'homologie. Il serait pourtant intéressant de distinguer les façons dont elle s'applique à ces deux phénomènes. L'image du bricolage en tant qu'improvisation dans les limites du recyclage s'appliquerait plutôt aux limites de l'analogie. Autrement dit dans quelle mesure un ichtyosaure doit rester un reptile facilement identifiable ? Une fois adopté le " bauplan " du reptile, le bricoleur voit se fermer devant lui beaucoup d'autres options. La convergence doit donc se construire avec les parties du reptile et porter avec elle la signature de son passé. Le bricolage est, par contre, la description adéquate du phénomène de l'homologie, c'est-à-dire de l'existence des segments, identiques du point de vue de leur information génétique, qui se trouvent en des lieux différents sur les chromosomes, où ils participent au " bauplan " des protéines lesquelles remplissent des fonctions divergentes.

2. FÉCONDITÉ

Un examen attentif de " Science citations index " (SCI) apprend que l'article de Jacob, paru en 1977[3] et en particulier les paragraphes consacrés à l'image qu'il proposait, sont cités en moyenne une fois par mois.

Je commencerai par l'aspect du recyclage qui a inspiré de nombreux travaux et je n'en mentionnerai que trois. Dans son article : " The function of the hereditary materials : biological catalyses reflect the cell's evolutionary history ", paru en 1986[4], B.M. Alberts propose le ribosome comme exemple du bricolage d'un appareil compliqué qui sert à la production des protéines à partir des restes d'un monde ancien, celui du temps où les molécules de l'A.R.N. servaient à la catalyse. Ce qui existe sert, même s'il faut le " rapetasser " en y ajoutant des protéines. Le deuxième exemple est l'article de Richard J. Goss : " The evolution of regeneration : adaptive of inherent ? ", paru en 1992[5]. L'auteur y trouve son chemin entre la position de Weismann, selon lequel la capacité de la régénération est un phénomène d'adaptation, et le scepticisme de Morgan. Goss opte pour la régénération inhérente et rappelle la métaphore proposée par Jacob. *Although there are many ways one could imagine to develop a replacement of a lost structure, nature does it the economical way by using pre-existing mechanisms encoded in the genes*[6]. Enfin, l'article de P.J.

---

3. *Ibid.*

4. B.M. Alberts, " The function of the hereditary materials : biological catalyses reflect the cell's evolutionary history ", *Amer. Zool.*, 26 (1986), 781-796.

5. Richard J. Goss, " The evolution of regeneration : adaptive or inherent ? ", *J. Theor. Biol.*, 159 (1992), 241-260.

6. *Id., Ibid.*, 249.

Keeling and W. Ford Doolittle : " Archaea : narrowing the gap between proka-ryotes and eukaryotes ", paru en 1995[7].

Il y a 30 ans nous pensions que la ligne de démarcation entre les bactéries et les autres organismes, était l'unique dichotomie de l'évolution. Nous savons à présent que les archaebacteria sont les plus proches parents des eucaryotes et nous reconnaissons trois branches de l'évolution. Les auteurs concluent que notre vision du monde a changé et qu'elle correspond mieux à l'image propo-sée par Jacob. *While cellular processes themselves differ between eukaryotes, eubacteria and archaebacteria, the components involved in carrying them seem seldom to have been purposefully built : Jacob's metaphor of evolution as tinkerer is as apt for molecules as for morphology*[8].

Je passe au deuxième aspect de l'activité combinatoire, à l'improvisation. Je me limiterai à trois travaux dont les sujets sont les sauts apparents dans l'évo-lution, mais dont les auteurs mettent en valeur le " comment " de leurs réalisa-tions. Je trouve le premier exemple dans les contributions de J. Piatigorsky et de ses collaborateurs, dont la première est l'article paru en 1992 : " Lens crys-talline, innovation associated with changes in gene-regulation "[9]. Les cristal-lins sont des protéines concentrées dans la lentille de l'oeil et assurent sa transparence. L'analyse génétique a révélé l'existence d'une homologie entre les cristallins, les protéines de choc thermique chez la drosophile d'une part, et les enzymes métaboliques tels que le glutathione S transférase (GST), d'autre part. En 1996, J. Piatogorski et son équipe[10] ont découvert que la perte de l'activité enzymatique de la famille des cristallins de type S, homologues à GST, concentrés dans les lentilles chez les céphalopodes, s'est réalisée gra-duellement grâce aux événement de duplication, mutation ainsi qu'aux séquen-ces d'insertion d'ordre et de longueur différentes. Le saut apparent de l'évolution des lentilles transparentes est le terme de petits pas sur la route du bricolage. Je trouve le deuxième exemple dans les travaux parus dans le sillage de l'article de Jacob, paru en 1993 : " Du répresseur à l'agrégulat "[11], dans lequel, tirant une conséquence de son image, il proposait le nom d'" agrégulat " pour la régulation de la transcription qui serait assurée par des agrégats de facteurs de transcription. De tels agrégulats permettent la création d'un large répertoire de systèmes régulateurs à partir d'un nombre limité de facteurs de transcription. La contribution de P. Santamaria, parue à la même

7. P.J. Keeling & W. Ford Doolittle, " Archaea : narrowing the gap between prokaryotes and eukaryotes ", *Proc. Natl. Acad. Sci.,* 92 (1995), 5761-5764.

8. *Id., Ibid.,* 5764.

9. L. Piatigorsky, " Lens cristallins. Innovation associated with changes in gene-regulation ", *The Journal of biological chemistry,* 267, 7 (1992), 4277-4280.

10. S.I. Tomarev, S. Chang, J. Piatigorsky, " Glutathione S - transferas and S - crystallins of cephalopods : evolution from active enzyme to lens refractrive proteines ", *Journal of molecular evolution,* 41, 6 (1996), 1048-1056.

11. F. Jacob, " Du répresseur à l'agrégulat ", *C.R. Acad. Sc. Paris, Sciences de la vie/Life sciences,* 316 (1993), 331-333.

date[12], étudiait les agrégulats qui fonctionnent au niveau de la différentiation cellulaire. L'auteur y traite des gènes du groupe Polycomb qui constitue un système combinatoire de gènes homéotiques chez la drosophile. Des changements dans la composition des éléments du système rendent compte aussi bien des phénotypes des mutants, que des étapes dans les processus évolutifs des diptères. Un troisième exemple enfin, est l'article de P. Rakic paru en 1995, dont le titre même exprime l'intuition que nous essayons d'illustrer : " A small step for the cell, a giant leap for mankind : a hypothesis of neocortical expansion during evolution "[13]. According to the proposed model, cortical expansion is the result of changes in proliferation kinetics that increase the number of radial columnar units, without changing the number of neurones within each unit significantly.

Voici donc pour recyclage et improvisation en tant qu'images inspiratrices. Enfin je voudrais souligner que l'image proposée par Jacob a aussi contribué à encourager l'étude des préadaptations. Les constructions et mécanismes que portent avec eux des organismes " patients ", représentent dans certaines circonstances un surcroît de bagage, mais ils peuvent se révéler, le cas échéant fonctionnels. G.L. Stebbis and D.L. Hartl ont reconnu leur dette à Jacob dans leur contribution parue en 1988 : " Comparative evolution : latent potentials for anagenetic advance "[14] ; ainsi E. Terzaghi and M. O'Hara en 1990 : " Microbial plasticity. The relevance to microbial ecology "[15]. L'image proposée par Jacob éclaire donc aussi bien l'interprétation des faits et des processus au niveau du génome, que l'évolution des organismes. Dans ce champ de recherche ainsi élargi, cette image a le mérite de proposer un chemin de milieu.

3. L'IMAGE DU JUSTE MILIEU

L'image proposée par Jacob propose un juste milieu entre les positions de Saint-Hilaire et de Cuvier. Le bricolage fait le lien entre le conservatisme des motifs et les possibilités qu'ouvre leur réorganisation. Comme disait Saint-Hilaire : " La nature utilise toujours les mêmes matériaux et son invention s'exprime dans la variation des formes "[16]. Mais aussi selon Cuvier, on peut identifier le noyau d'organisation qui, une fois existant, contraint la suite de

---

12. P. Santamaria, " Evolution and aggregulates : role of the Polycomb - Group genes of Drosophila ", *C.R. Acad. Sc. Paris, Sciences de la vie/Life sciences,* 316 (1993), 1200-1206.

13. P. Rakic, " A small step for the cell, a giant leap for mankind : a hypothesis of neocortical expansion during evolution ", *Tins.,* 18, 9 (1995), 383-388.

14. G.L. Stebbis & D.L. Hartl, " Comparative evolution : latent potentiels for anagenetic advance ", *Proc. Natl. Acad. Sci. (U.S.A.),* 85 (1988), 5141-5145.

15. E. Terzaghi & M. O'Hara, " Microbial plasticity. The relevance to microbial ecology ", *Advances in Microbial Ecology,* 11 (1990), 431-460.

16. E. Geoffroy Saint-Hilaire, *Philosophie Anatomique,* Paris, 1818, 18-19.

l'organisation[17]. De même, encore selon Cuvier, les différentes constructions se révéleraient fonctionnelles selon les occasions géophysiques des environnements.

L'image du bricolage a eu aussi le mérite de souligner la coexistence de quatre notions-clés qui doivent guider la recherche à chacun des niveaux de l'organisation du vivant : conservation, variabilité, redondance et parcimonie. Ainsi est indiqué le passage entre les " deux niveaux " de l'évolution, l'un moléculaire, l'autre anatomique, et le " grand divorce " est raccommodé[18]. La métaphore de l'improvisation dans les limites du recyclage indique aussi le juste milieu entre la conception de l'évolution comprise comme une série de " sauts " étonnants selon l'alternative proposée par N. Eldredge and S.J. Gould en 1972, " An alternative to phyletic gradualism "[19], et celle selon laquelle on devrait chercher une explication " adaptative " pour chacune des " grandes transitions de l'évolution ", selon le titre du livre de J.M. Smith et E. Szathmary, paru en 1995[20].

En tant que chemin de milieu, l'image proposée par Jacob a été aussi une source d'inspiration pour des programmes apparemment contradictoires. Ainsi, d'une part l'étude du pouvoir explicatif des contraintes phylogénétiques : pourrait-on les utiliser pour rendre compte de l'absence de certains phénomènes pourtant *prévisibles* dans le développement ? Ce programme de recherche fut proposé par M.C. McKitrick qui cite Jacob dans son article paru en 1993 : " Phylogenetic constraint in evolutionary theory : has it explanatory power ? "[21]. D'autre part l'on met en lumière l'*imprévisible* que représente pour nous autres humains, la manière de faire du bricolage dans l'évolution. Pour F. Crick, tel qu'il a expliqué dans son article paru en 1989[22], cette manière de faire devrait plutôt atténuer l'inclination à expliquer le fonctionnement du cerveau en se référant aux modèles proposés par le fonctionnement des ordinateurs. Je cite, en conclusion, F. Crick : " ...evolution is a tinkerer ...naturally it is constrained by both chemistry and physics, but this does not necessarily mean that its mechanism will embody deep general principles... It may prefer a series of slick tricks to achieve its aim "[23].

17. G. Cuvier, *Leçon d'Anatomie*, 2ᵉ éd., Paris, 1835, vol. I, 59.

18. M.C. King & A.C. Wilson, " Evolution at two levels in humans and chimpanzees ", *Science*, 188 (1975), 107-116.

19. N. Eldredge & S.J. Gould, " Punctuated Equilibria : an alternative to phyletic gradualism ", *Models in Paleobiology*, C.T.J.M., Schopf ed., 1972.

20. J.M. Smith & E. Szathmary, *The major Transitions in Evolution*, Oxford, 1995.

21. M.C. McKitrick, " Phylogenetic constraint in evolutionary theory : has it explanatory power ? ", *Annu. Rev. Ecol. Syst.*, 24 (1993), 307-330.

22. F. Crick, " The recent excitement about neural networks ", *Nature*, 337, 12 (1989), 129-132.

23. *Id., Ibid.*, 132.

NOTE

On se reportera au récent article de D. Duboule et A.S. Wilkins, " The evolution of 'bricolage' ", *Trends in Genetics*, 14 (1998), 54-59.

Making Sense of Life[1]

Evelyn Fox Keller

With the extraordinary developments of the last 20 years in the molecular analysis of developmental genetics, Developmental Biology has undergone a renaissance ; indeed, it is not uncommon to hear claims that, finally, we have arrived at an understanding of the " basic principles " of development (see, e.g., Wolpert, 1994 : 571). Others, however, see things differently. Adam Wilkins, e.g., writes that, " The added wealth of molecular description in the last two decades has certainly deepened our understanding, but the underlying mechanisms still remain largely hidden. …The fundamental problems of understanding developmental change are …to a large extent still very much with us " (1993 : 16).

For many, it is the very success of molecular analysis that has elicited a recognition of the explanatory limits of the genetic paradigm, i.e., of traditional notions of genes (or genetic programs) as ultimate causes of (or instructions for) development. In this paper, I examine such notions and juxtapose them with an alternative perspective, namely one which takes as its starting point not genes, but rather, the manifest robustness of the developmental process. These two perspectives lead to radically different questions, and to radically different criteria of explanation. The role of redundancy in the two frameworks is illustrative of the differences : in one, redundancy appears as an unexpected and possibly inexplicable problem ; in the other, just because of the role of redundancy in establishing robustness, as a first and necessary ingredient of an acceptable explanation.

Elsewhere, I have written of the rise of a " discourse of gene action " among American geneticists in the 1920s and 1930s as a way of " accounting " for development when little was known about either what genes were or how they worked (1995). Enormous changes have occurred since those early days, in

---

1. Taken from a longer essay under the same title, to appear in Maienschein and Creath, (forthcoming).

biology as elsewhere. As a result of the dramatic achievements of molecular biology, we now have a vast amount of information about the nature of genes and about their involvement in specific developmental phenomena. In conjunction with these developments, talk of " gene action " has largely given way to " gene activation ". Nevertheless, crucial elements of that earlier discourse still persist, most notably in the widespread notion that the development of a trait or function has been explained when the gene or genes " for " that trait or function has been identified, and, correlatively, that development as a whole can be explained by enumerating the genes responsible. Indeed, this notion underlies not only genetic explanations of development, but also the very logic of much if not most of genetic research.

Current research however provides a number of challenges to this logic. One such challenge emerges with the recognition that many genes assumed necessary for particular traits are only necessary in particular genetic contexts. Tacit acknowledgement of this limitation routinely appears in the literature with such qualifications as, e.g., gene x " is usually necessary for ", " is involved in ", or more simply " has something to do with ", the appearance of the trait in question (see Keller (forthcoming) for further discussion).

A complementary challenge to the genetic paradigm arises from what might be thought of as the obverse phenomenon, namely the appearance of null mutants which show no phenotypic effect even when both alleles of the gene in question are dysfunctional. These mutants are called " null " because of their failure to produce particular enzymes which are produced by the wild type, and which are assumed to serve an important function because of the extent to which they have been conserved over evolution. Such " null " mutants came as a big surprise when they were first identified in the early 1970s, and they were generally regarded as quite mystifying. Over the last twenty years, they have become commonplace, if somewhat mystifying. Thanks largely to the acquisition of new molecular techniques for targeted mutagenesis, evidence for such phenomena has accumulated exponentially and the consensus among researchers in the field is that they clearly indicate the existence of functionally redundant genetic pathways. Today, functional redundancy — on the level of genetic transcription, transcriptional activation, genetic pathways, and of intercellular interactions — has emerged as a prominent feature of developmental organization in complex organisms.

Redundancy is a staple of engineering, but it has been said to " strike fear in the heart of geneticists " (Brenner et al., 1991), and for good reason : it not only reveals limits to the value of mutation screening (its core technique) in probing developmental dynamics — limits that had earlier been only implicit[2] ; it is also seen as a threat to the entire explanatory framework of the

2. Such limits were inferable, e.g., from the ubiquitous phenomenon of variable penetrance (see, e.g., Wilkins, 213).

genetic paradigm. As Diethard Tautz writes, " Though the geneticist will often be unable to say exactly how a certain mutation causes a certain phenotype, ...he must maintain that single and direct causal relationships exist. This genetic paradigm is at the basis of all systematic mutagenesis experiments,... [But] even the best paradigm eventually meets a crisis. Such a crisis is imminent " (1992 : 263).

These two challenges to the assumption of genes as the causal units of development — one a consequence of inhibition (or repression), the other of redundancy — are both features of the complex regulatory networks in which genes are now found to be enmeshed. Here is how the Nobel Laureate J. Michael Bishop describes the problem : " What at first appeared as simple linear arrays of switches has now emerged as an elaborate network with hundreds (if not thousands) of nodal points. Attempts to trace a signal through the circuitry soon become lost in a welter of crosstalk and feedback... It seems unlikely that [our] efforts will be fully effective unless a global view of the molecular circuitry can be achieved " (1995 : 1617).

But what would such a global view entail ? What kind of explanation might it be expected to yield ? And finally, what, if any, alternatives to the present explanatory framework of genetics might be available for furthering our thinking about development ?

Understood in the most general terms as the capacity of developing organisms to compensate for disruptions, the robustness of development has been a primary concern for embryologists every since Driesch's initial observation of regulation (1891). In an effort to explain instances of robustness in later stages of development, a specific reference to engineering principles surfaced early under the term " double assurance " (see, e.g., Spemann, 1938). But it was C.H. Waddington who, in the 1940s, first attempted to relate observations of developmental stability to genetics, and to outline an explanatory framework in terms of interacting " gene-protein systems ". Waddington coined the term " creode " to denote the developmental pathway, or trajectory, that resulted from the interaction of organized genetic systems and environmental effects, and the term " developmental canalization " to denote the progressive stability of these trajectories. Although his particular concern here was with the stabilization of cell differentiation, rather than with the overall robustness of embryogenesis (indeed, his metaphor of alternative pathways breaks down for embryogenesis as a whole where the key point is the (relative) fixedness of the final state a zygote will reach), his picture of the epigenetic landscape was useful in drawing attention to stabilized or buffered pathways of change as a central feature of development (see, e.g., Waddington, 1957 ; Gilbert, 1991). An explanation of such " developmental canalization ", he argued, required that the discrete and separate entities of classical genetics would be displaced by collections of genes which could " lock in " development through their interactions (Waddington, 1948).

Today, Waddington seems to be enjoying something of a revival, and it has been suggested that his work on canalization constitutes " a premature discovery " (Wilkins, 1997 : 257). Yet, in his own time it held little interest for most of his colleagues, and in the decades that followed, it had scarcely any impact on the course of developmental research. Why is that ? One might have thought that the stability (or fidelity) of the developmental process is its most conspicuous feature ; indeed as its most basic feature, even as the essential pre-condition of evolution, for without an internal capacity to withstand the inevi-table vicissitudes of ordinary development, organisms of a particular genotype would not persist long enough for selection to act upon them[3]. It is precisely the remarkable constancy of the developmental process — its resilience in the face of so much cytoplasmic, environmental, and even genetic variation — that had led earlier thinkers to think of it as inherently goal-oriented, as internally directed towards a fixed final state. Indeed, it was just this feature of constancy that the original notion of a " genetic program " was intended to capture. But however construed, the only account for stability it was ever able to provide was limited to whatever structural stability could be attributed to the DNA itself, coupled with the degree of fidelity that could be attributed to its capacity for spontaneous " self-replication " and transmission of " instructions ". As it turns out, the answer is not much. The structural stability of the chromosome is not provided by the DNA itself but by the proteins with which it interacts, and the degree of fidelity one actually observes in both copying and transmis-sion is now known to depend on the presence of an elaborate machinery of proof-reading and repair[4]. In fact, the constancy of development depends on its overall stability, and that striking feature remains effectively unaccounted for.

A focus on developmental stability points to questions (and hence towards explanations) radically different from any that can even be asked within the genetic paradigm. In effect, by its dependence on mutational analysis, the latter seeks to explain development by asking what causes it to fail (or go astray) ; the assumption is that the causes of normal development can be inferred by logical subtraction — i.e., by an enumeration of all those genes that can be identified by their phenotypic failure. By contrast, a focus on developmental stability leads one to ask, what is required to make it work, what is it that endows the developmental process with such reliability ? This distinction is familiar to any engineer attempting to formulate design principles for complex systems predicated on the reliability of performance (e.g., the design of air-planes which can be reasonably certain of reaching their destination despite

3. This view suggests that the role of contingency in development is the obverse of its role in evolution — where contingency is said to be the definitive characteristic of evolution, it is resis-tance to contingency that could be said to be definitive of development.

4. As Alberts *et al.* write, " genetic information can be stored stably in DNA sequences only because a large variety of DNA repair enzymes continuously scan the DNA and replace the dam-aged molecules " (1989 : 227).

vicissitudes of weather, air traffic, *etc.*). Engineers have long understood the crucial importance of redundancy in guaranteeing reliability. In other words, their aims (or concerns) led them early on to just that feature of design that has been most inaccessible to the traditional techniques of genetic analysis.

It is little surprise, therefore, that evidence for widespread redundancy in developmental systems has prompted talk of a crisis in the traditional paradigm of genetics. From that perspective, redundancy is not only technically opaque, but it also doesn't seem to make sense. If genes are taken as the units of selection, how could redundancy have evolved ? Indeed, it is just this need to make evolutionary sense of redundancy that leads Diethard Tautz to invoke a familiar lesson from information theory, namely, that fidelity in the transmission of information requires redundancy, and to suggest an obvious analogue for living systems. He writes " The formation of an adult organism can be seen as the transmission of information which is laid down in the egg and its genome. ...At each [developmental step] there is a potential loss of information and the developing organism has to safeguard itself against this loss. This is, of course, a good basis for selection pressure to evolve redundancies. This selection pressure need not be very high, since even a small effect on the probability of successful completion of embryogenesis would directly be reflected in the probability of survival of the offspring. ...Thus, the evolution of redundant regulatory pathways may be seen as a logical consequence of the evolution of complex metazoan life " (1992 : 264).

Or perhaps as the logical precondition of metazoan life. Either way, the significant point is that the need to make sense of redundancy has led Tautz and others to just the kind of global view that Bishop advocates. In this view, the unit of selection is not the gene, but the life cycle itself.

REFERENCES

B. Alberts *et al.*, *The Molecular Biology of the Cell*, Garland Press, 1989.
J. Michael Bishop, " Through a Glass Darkly ", *Science*, 267 (1995), 16-17.
Sydney Brenner *et al.*, " Genes and Development : Molecular and Logical Themes ", *Genetics*, 126 (1990), 479-486.
Scott Gilbert, " Induction and the Origins of Developmental Genetics ", in S. Gilbert (eds), *A Conceptual History of Modern Embryology*, Johns Hopkins Univ. Press, 1991, 181-205.
François Jacob, *The Logic of Life,* 1970 (English paperback edition, Vantage Books, 1976).
Evelyn Fox Keller, " Making Sense of Life : Explanation in Developmental Biology ", in Maienschein and Creath, forthcoming.
Evelyn Fox Keller, *Refiguring Life*, Col. U. Press, 1995.
Hans Spemann, *Embryonic Development and Induction*, Yale Univ. Press, 1938.

Diethard Tautz, " Redundancies, Development and the Flow of Information ",
    Bio Essays, 14 (4) (1992), 263-266.

C.H. Waddington, " The Genetic Control of Development ", *Symp. Soc. Exp.
    Biol.*, 11 (Acad. Press, 1948).

C.H. Waddington, *Strategy of the Genes*, London, Allen & Unwin, 1957.

Adam Wilkins, *Genetic Analysis of Animal Development*, NY, Wiley-Liss,
    1992.

Adam, Wilkins, " Canalization : a molecular genetic perspective ", *BioEssays*,
    19 (3) (1997), 257-262.

Lewis Wolpert, " Do We Understand Development ? ", *Science*, 266 (28 Octo-
    ber 1994), 271-272.

CONTRIBUTORS

Richard H. BEYLER
Department of History
Portland State University
Portland, OR (USA)

Richard M. BURIAN
Science Study Center
Virginia Polytechnic University
Blacksburg, VA (USA)

Johannes BÜTTNER
University Medical School
Isernhagen (Germany)

Jean-Claude DUPONT
Département de Philosophie
Université de Picardie
Amiens (France)

Raphael FALK
Department of Genetics
The Hebrew University
Jerusalem (Israël)

Jean-Louis FISCHER
Centre Alexandre Koyré
Museum d'histoire naturelle
Paris (France)

Evelyn FOX KELLER
Program in Science
Technology and Society
Massachusetts Institute of
Technology
Cambridge, MA (USA)

Charles GALPERIN
CRATS-URA 1743 - CNRS
Université Charles de Gaulle
Villeneuve d'Ascq (France)

Jean GAYON
Université Paris 7 Denis Diderot
Institut Universitaire de France
Paris (France)

Scott F. GILBERT
Swarthmore College
Department of Biology
Swarthmore, Pennsylvania (USA)

Vera GUTINA
Inst. for History of Science
and Technology
Moscow (Russia)

Brigitte HOPPE
Inst. für Geschichte der
Naturwissenschaften der
Universität München
München (Germany)

Jan JANKO
Inst. for Fundamental Learning
Charles University
Praha (Czech Republic)

Hervé LE GUYADER
Université de Paris Sud
Laboratoire de Biologie Cellulaire
Orsay (France)

Claudia LENZNER
Institut für Biochemie der Charité
Humbold Universität zu Berlin
Berlin (Germany)

Alfred NEUBAUER
Berlin (Germany)

Nadine PEYRIERAS
Biologie Moléculaire
du Développement
INSERM U 368
École Normale Supérieure
Paris (France)

Hans-Jörg RHEINBERGER
Max Planck Institute
for the History of Science
Berlin (Germany)

Stéphane SCHMITT
CEHM
Strasbourg (France)

Christiane SINDING
INSERM
CNRS/UPR 1524

Hôpital Saint-Vincent de Paul
Paris (France)

Denis THIEFFRY
Max Planck Inst. for History of
Science
Berlin (Germany)

Patrick TRIADOU
Laboratoire central d'hématologie
Tour Pasteur
Hôpital Necker
Paris (France)

Avital WOHLMAN
The Hebrew University
Department of Philosophy
Jerusalem (Israël)